Math Mammoth
Add & Subtract 2-A

By Maria Miller

Contents

Introduction

Math Mammoth Add & Subtract 2-A deals with two main themes:

- strategies for adding and subtracting within 0-20; such as adding just one more, a trick with nine and eight, and subtracting using addition;

- memorizing the basic addition and subtraction facts of single-digit numbers with an answer between 10 and 18.

While focusing on addition and subtraction techniques, the lessons also include many word problems.

The goal is to memorize the facts, or at least become so fluent with them that an outsider cannot tell if the child remembers the answer or uses some mental math strategy to get the answer.

Some children will accomplish this quicker and need less practice, whereas others will take longer. Thus, don't assign all the exercises in the book by default. Use your judgment, and try to match the amount of exercises to your child's need. The ones that don't get assigned can be used later for review. You can also use games to reinforce the facts, and in place of some of the exercises in the book (a list of games is provided below).

Learning addition and subtraction facts is quite important for later study. For example, regrouping in addition and subtraction (carrying/borrowing) requires the ability to recall the basic facts efficiently and fluently.

We will start the book with a few mental math strategies (Add Using "Just One More" and A "Trick" with Nine and Eight). The lesson Adding within 20 reviews those strategies and gives more practice. These initial lessons don't yet involve actual memorization techniques.

The lesson *Subtract to Ten* explains another basic strategy, and has to do with subtracting in parts. For example, to do $13 - 5$, one can subtract 5 in two parts: first do $13 - 3$, which equals 10, and then subtract the rest, or 2 more. Memorizing the subtraction facts will be more efficient, but I want children to understand this strategy, because it is useful in many other situations also.

Then we review how to complete the next whole ten, which is an important concept. An example of this concept is the question: What number do you add to 23 to get 30? As an equation, we write: $23 + \_\_ = 30$.

In the next lesson, we study sums that go over ten, doing these sums in two parts. For example, in the sum $9 + 7$, the child first completes 10 by adding $9 + 1$. Then, the child adds the rest, or 6, to 10. Learning this prepares the child for addition facts where the sum is more than 10.

The next lessons, *Adding with 9*, *Adding with 8*, *Adding with 7*, and *Adding with 6*, provide lots of practice for learning and memorizing the basic addition facts. There are 20 such facts:

$9 + 2$ till $9 + 9$: 8 facts
$8 + 3$ till $8 + 8$: 6 facts
$7 + 4$ till $7 + 7$: 4 facts
$6 + 5$ till $6 + 6$: 2 facts

The last part of the book includes various lessons titled *Number Rainbows* and *Fact Families with....* These give lots of practice and reinforcement for the basic addition and subtraction facts, emphasizing the connection between addition and subtraction as a strategy for subtraction facts.

You can find videos that match these particular lessons at **https://www.mathmammoth.com/videos**. Choose 2nd grade, then the section for addition and subtraction facts.

> *I wish you success in teaching math!*
>
> *Maria Miller, the author*

Games and Activities

12 Out (or *11 Out, 13 Out, 14 Out*)

You need: A deck of number cards, or regular playing cards. The values of the face cards are Jack = 11, Queen = 12, King = 13.

Preparation: Choose a target sum, such as 12. The game works best for target sums 14 or less. Deal seven cards to each player. Place the rest face down in a pile in the middle of the table.

Game play: At your turn, first take one card from the pile. Then try to find pairs of cards in your hand that add up to 12, and discard any such pairs. Discard the card 12 (queen) also if you have it. If you cannot find any such pairs, ask for any one card you want (such as 7) from the player to your right (as in "Go Fish"). That player, if they have it, must give it, and you will then discard the pair that makes 12. Then it is the next player's turn. The player who first discards all the cards from their hand is the winner.

Variations:
* Deal more than seven cards.
* Instead of 12, players discard cards that add up to 12, 13, or 14.

Addition (or Subtraction) Challenge

You need: A standard deck of playing cards from which you remove the face cards. For the subtraction challenge, include the face cards also (Jack = 11, Queen = 12, King = 13).

Game Play: In each round, each player is dealt two cards face up, and has to calculate their sum or difference (add/subtract). The player with the highest sum or difference gets all the cards from the other players. After enough rounds have been played to use all of the cards, the player with the most cards wins. If two or more players have the same sum, then those players get an additional two cards and use those to resolve the tie.

Number Bonds in the Pond

You need: A standard deck (or several) of playing cards or number cards. The values of the face cards are Jack = 11, Queen = 12, King = 13.

Preparation: Choose a target sum for the game. If the target sum is 12, make a deck of cards consisting of numbers 1 through 11. If the target sum is 11, make a deck of numbers 1-10. And so on. (The deck always consists of numbers that are from 1 through $X - 1$ where X is the target sum.) Place a target number card face up between the players, and spread out the rest of the cards face down, like a pond, between the players.

Game play: At your turn, if you don't have any cards in your hand, take <u>two</u> cards from the pond. If you do, take <u>one</u> card from the pond. Now check if any two cards in your hand add up to the target number. If so, put those cards away to your personal pile. If not, it is the next player's turn. The game ends when there are no more cards in the pond. The winner is the person with the most cards in their personal pile.

Variation: Allow three cards/numbers to be added to reach the target number.

Note: Depending on the number of players, you may need several decks of cards for the pond.

Get Out of My House

You need: A deck of playing cards or number cards from 3 to 10.

Preparation: On a shared piece of paper, draw boxes (houses) numbered from 6 to 20. This works best as a two-player game, and each player needs seven tokens that are distinct from the other player's tokens. Place the deck of cards in the middle, cards face down.

Game play: During a turn, a player takes two cards from the deck, adds them, and then puts their token in a house with fewer than three of the opponent's tokens. If the house contains one or two of the opponent's tokens, those tokens are given back to the opponent and the player says "Get out of my house." The first player to place all their tokens in houses wins.

Variation: Allow subtraction and/or multiplication to be used, along with addition.

This game is adapted from https://www.earlyfamilymath.org and published here with permission.

Games and Activities at Math Mammoth Practice Zone

Single-Digit Addition
Simple practice of addition facts with single-digit addends.
https://www.mathmammoth.com/practice/addition-single-digit#questions=10&toe=18&pt=general

Hidden Picture Addition Game
Use a number range of 2 to 9 to specifically practice basic addition facts.
https://www.mathmammoth.com/practice/mystery-picture

7 Up Card Game
You will see seven cards dealt face up. Simply choose any two cards that make 10 (or your chosen sum) to discard. When there are no cards that make that sum, click the deck to deal more cards. For this chapter, choose sums of 11, 12, 13, and 14.
https://www.mathmammoth.com/practice/seven-up

Fact Families
Choose which fact family or families to practice, and the program will give you addition and subtraction problems from those, including with missing numbers. For this chapter, choose fact families with 11, 12, 13, 14, and 15.
https://www.mathmammoth.com/practice/fact-families

Mathy's Berry Picking Adventure
Join Mathy, our mammoth mascot, on his berry-picking adventure, and practice your basic addition or subtraction facts!
https://www.mathmammoth.com/practice/mathy-berries#mode=addition-single&duration=2m

https://www.mathmammoth.com/practice/mathy-berries#mode=sub-20&duration=2m

Bingo
Simply click on the right answer in the grid, and it will be colored green. Once you get five in a row, a column, or diagonally, and bingo, you win! For this chapter, choose Addition (Single-Digit) or Subtraction (Under 20).
https://www.mathmammoth.com/practice/bingo

Fruity Math
Click the fruit with the correct answer and try to get as many points as you can within two minutes. The first link below is for addition facts, the second one for subtraction within 0-18.
https://www.mathmammoth.com/practice/fruity-math#op=addition&duration=120&mode=manual&config=2,9x1__3,9x1&max-sum=1000

https://www.mathmammoth.com/practice/fruity-math#op=subtraction&duration=30&mode=manual&config=11,18x1__2,9x1&allow-neg=0

Helpful Resources on the Internet

We have compiled a list of external Internet resources that match the topics in this book. This list of links includes web pages that offer:

- **online practice** for concepts;

- online **games**, or occasionally, printable games;

- **animations** and interactive **illustrations** of math concepts;

- **articles** that teach a math concept.

We heartily recommend you take a look at the list. Many of our customers love using these resources to supplement the bookwork. You can use the resources as you see fit for extra practice, to illustrate a concept better, and even just for some fun. Enjoy!

https://l.mathmammoth.com/blue/addsubtract2a

SCAN ME

Add Using "Just One More"

Do you remember the numbers that add up to 10 ("the sums of 10")?
There are 9 and 1, and what others? List them now.

JUST ONE MORE than a sum of 10:			
$8 + \boxed{2} = 10$ $8 + \boxed{3} = 11$	8 + 3 is JUST ONE MORE than 8 + 2, so the answer is also just one more.	$\boxed{5} + 5 = 10$ $\boxed{6} + 5 = 11$	6 + 5 is JUST ONE MORE than 5 + 5, so the answer is also just one more.

1. Change the underlined number to be JUST ONE MORE. The answer changes, too!

a. $8 + \underline{2} = 10$ $8 + \underline{3} = \underline{}$	b. $4 + \underline{6} = 10$ $4 + \underline{} = \underline{}$	c. $\underline{7} + 3 = 10$ $\underline{} + 3 = \underline{}$
d. $\underline{1} + 9 = 10$ $\underline{} + 9 = \underline{}$	e. $5 + \underline{5} = 10$ $5 + \underline{} = \underline{}$	f. $\underline{4} + 4 = 8$ $\underline{} + 4 = \underline{}$

2. Find the missing numbers.

a. $7 + \boxed{} = 10$ $7 + \boxed{} = 11$	b. $8 + \boxed{} = 10$ $8 + \boxed{} = 11$	c. $6 + \boxed{} = 10$ $6 + \boxed{} = 11$
d. $5 + \boxed{} = 11$	e. $9 + \boxed{} = 11$	f. $3 + \boxed{} = 11$

3. Add. Think of JUST ONE MORE. Color the problems where you use that idea!

a.	b.	c.	d.
$7 + 2 = \underline{}$	$5 + 6 = \underline{}$	$4 + 6 = \underline{}$	$2 + 9 = \underline{}$
$3 + 8 = \underline{}$	$3 + 4 = \underline{}$	$2 + 8 = \underline{}$	$5 + 4 = \underline{}$
$5 + 5 = \underline{}$	$6 + 4 = \underline{}$	$7 + 4 = \underline{}$	$3 + 7 = \underline{}$

> The **double** of something means twice (two times) that thing.
>
> For example, "double four" means 4 and 4. So double 4 is 8. How much is double 3? Double 5?
>
> Double six, or $6 + 6$, is 12.
>
> We can use that to find $6 + 7$. It is JUST ONE MORE! It is 13.

4. On the right you see a **doubles chart**. You can use it for the addition problems below. Think of "JUST ONE MORE!"

a. $7 + 6 =$ _____	**b.** $7 + 7 =$ _____	**c.** $9 + 8 =$ _____	$5 + 5 = 10$
d. $8 + 8 =$ _____	**e.** $5 + 6 =$ _____	**f.** $9 + 10 =$ _____	$6 + 6 = 12$
g. $7 + 8 =$ _____	**h.** $9 + 9 =$ _____	**i.** $6 + 5 =$ _____	$7 + 7 = 14$
j. $8 + 9 =$ _____	**k.** $6 + 7 =$ _____	**l.** $8 + 7 =$ _____	$8 + 8 = 16$
			$9 + 9 = 18$

5. Solve the word problems.

a. Joe bought a package of 12 balloons. He gave three to Sam, two to his sister and five to Jane. How many balloons did he give away?

How many balloons does Joe have left?

b. Marsha found seven uniforms for the softball teams in one box, and six more uniforms in another box. How many uniforms did Marsha find?

c. Three of the uniforms Marsha found were clean, but she had to wash the rest. How many uniforms did Marsha have to wash?

d. Eight girls and five boys came to play softball. How many more girls came than boys?

e. Did Marsha have enough uniforms for the boys and girls who came to play softball? If not, how many more uniforms does she need? If so, how many uniforms were left over?

A "Trick" with Nine and Eight

A "trick" with nine

Imagine that 9 wants to be ten! It's not happy—
it wants to become a full TEN!

So, nine asks the other number (this time, 7) to
give him some in order to make himself to be a ten.

Seven says, "OK," gives one to 9, and has only
six left for himself.

In the end, we have 10 and 6. We get 16.

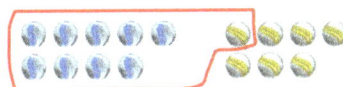

$$9 \quad + \quad 7$$

$$\downarrow \qquad\qquad \downarrow$$

$$10 \qquad\qquad 6 \quad = 16$$

We can also show the same thing this way →

Notice: it will also work if the second
number is 9. Why? Because you can add
in any order. $5 + 9$ is the same as $9 + 5$.

$$9 + 7$$
$$\quad | \ \backslash$$
$$9 + 1 + 6$$
$$10 + 6 = 16$$

---

1. Circle all of the blue marbles and some of the yellow ones so that you get a ten. Add.

a. $\quad 9 \quad + \quad 6$ $10 + \underline{\;5\;} = \underline{\hspace{2cm}}$	**b.** $\quad 9 \quad + \quad 4$ $10 + \underline{\hspace{1cm}} = \underline{\hspace{2cm}}$
c. $\quad 9 \quad + \quad 3$ $10 + \underline{\hspace{1cm}} = \underline{\hspace{2cm}}$	**d.** $\quad 9 \quad + \quad 5$ $10 + \underline{\hspace{1cm}} = \underline{\hspace{2cm}}$

2. Fill in the blanks. Imagine that nine wants to become a ten.

a. $\quad 9 + 8$ $\quad\quad / \quad \backslash$ $9 + \underline{\hspace{1cm}} + \underline{\;7\;}$ $10 + \underline{\hspace{1cm}} = \underline{\hspace{1cm}}$	**b.** $\quad 9 + 7$ $\quad\quad / \quad \backslash$ $9 + \underline{\hspace{1cm}} + \underline{\hspace{1cm}}$ $10 + \underline{\hspace{1cm}} = \underline{\hspace{1cm}}$	**c.** $\quad 9 + 9$ $\quad\quad / \quad \backslash$ $9 + \underline{\hspace{1cm}} + \underline{\hspace{1cm}}$ $10 + \underline{\hspace{1cm}} = \underline{\hspace{1cm}}$

11

A "trick" with eight

Imagine that 8 wants to be ten! It's not happy—it wants to become a full TEN!

So, eight asks the other number (this time, 5) to give him some in order to make himself to be a ten.

Five says, "OK," gives two to 8, and has only three left for himself.

In the end, we have 10 and 3. We get 13.

$$8 \quad + \quad 5$$
$$\downarrow \qquad \qquad \downarrow$$
$$10 \qquad \qquad 3 \quad = 13$$

We can also show the same thing this way:

$$8 + 5$$
$$\diagup \quad \diagdown$$
$$8 + 2 + 3$$
$$10 + 3 = 13$$

3. Circle all of the blue marbles and some of the yellow ones so that you get a ten. Add.

a. 8   +   6

   10 + _____ = _____

b. 8   +   7

   10 + _____ = _____

c. 8   +   3

   10 + _____ = _____

d. 8   +   4

   10 + _____ = _____

4. Fill in the blanks. Imagine that eight wants to become a ten.

a. 8 + 8
      /   \
8 + _2_ + _____
   10 + ____ = ____

b. 8 + 5
      /   \
8 + _____ + _____
   10 + ____ = ____

c. 8 + 7
      /   \
8 + _____ + _____
   10 + ____ = ____

5. Right or not? Cross out the additions that are *false* (not correct).

   a. $6 + 6 = 13$ **b.** $7 + 8 = 15$ **c.** $9 + 6 = 15$ **d.** $9 + 7 = 17$

6. Solve.

a. A basket has nine apples in it. Alice ate two, and her brother ate one. How many apples are left?	**b.** Jeremy picked up nine apples that had fallen under an apple tree. Then he picked up six more under another tree. How many apples does Jeremy have now?
c. Alice picked 7 flowers and Jeremy picked 9. How many more flowers did Jeremy pick? How many flowers did the children have together?	**d.** Jeremy put toy cars end-to-end. One car was 5 cm long, another was 5 cm also, and the third car was 4 cm long. How long was Jeremy's train of cars?

7. Write a number inside the balloon so that the numbers in the balloon make a ten. Add.

a.	**b.**	**c.**
$7 + \underline{3} + 5 = \underline{15}$	$9 + \underline{} + 2 = \underline{}$	$7 + \underline{} + 5 = \underline{}$
d.	**e.**	**f.**
$6 + \underline{} + 6 = \underline{}$	$8 + \underline{} + 4 = \underline{}$	$5 + \underline{} + 8 = \underline{}$

8. Add. Think how the nine or the eight wants to be ten! If the *second* number is 8 or 9, turn the addition around. You can add the numbers in the other order, 8 or 9 first.

a. $8 + 6 = \underline{}$ b. $6 + 9 = \underline{}$ c. $9 + 4 = \underline{}$

d. $4 + 8 = \underline{}$ e. $8 + 7 = \underline{}$ f. $9 + 9 = \underline{}$

g. $9 + 5 = \underline{}$ h. $8 + 8 = \underline{}$ i. $3 + 8 = \underline{}$

What number goes in the shape?   Puzzle Corner

a. $\triangle + 8 = 16$ b. $\triangle + 9 = 15$ c. $\triangle + 2 + 7 = 13$

13

Adding Within 20

You have learned many things to help you add when the sum (the answer) is more than 10. Let's review them:

1. The trick with nine and eight. $9 + 6 = ?$ Think of nine wanting to be ten, and so six gives one to nine. Then, the addition becomes $10 + 5$, which is 15.	2. Just one more than an addition you know. For example, $3 + 7 = 10$, so $3 + 8$ must be just one more, or 11.

3. The doubles chart:	4. Just one more than a double: $7 + 8$ is just one more than $7 + 7$. Since $7 + 7$ is 14, then $7 + 8$ must be 15.

$5 + 5 = 10$

$6 + 6 = 12$

$7 + 7 = 14$

$8 + 8 = 16$

$9 + 9 = 18$

1. Write here additions that you can solve using the idea "just one more" than a double.

a. $5 + 5 = 10$	**b.** $6 + 6 = 12$	**c.** $7 + 7 = 14$
$\underline{\ 5\ } + \underline{\ 6\ } = 11$ and $\underline{\ 6\ } + \underline{\ 5\ } = 11$	$\underline{\ \ \ } + \underline{\ \ \ } = 13$ and $\underline{\ \ \ } + \underline{\ \ \ } = 13$	$\underline{\ \ \ } + \underline{\ \ \ } = \underline{\ \ \ }$ and $\underline{\ \ \ } + \underline{\ \ \ } = \underline{\ \ \ }$
d. $8 + 8 = 16$	**e.** $9 + 9 = 18$	**f.** $10 + 10 = 20$
$\underline{\ \ \ } + \underline{\ \ \ } = 17$ and $\underline{\ \ \ } + \underline{\ \ \ } = 17$	$\underline{\ \ \ } + \underline{\ \ \ } = 19$ and $\underline{\ \ \ } + \underline{\ \ \ } = \underline{\ \ \ }$	$\underline{\ \ \ } + \underline{\ \ \ } = \underline{\ \ \ }$ and $\underline{\ \ \ } + \underline{\ \ \ } = \underline{\ \ \ }$

2. Add. Use the trick with nine.

a. $9 + 8 = \underline{\ \ \ }$	**b.** $3 + 9 = \underline{\ \ \ }$	**c.** $9 + 5 = \underline{\ \ \ }$	**d.** $6 + 9 = \underline{\ \ \ }$

3. For each sum with 10 write another that is "just one more."

a. 1 + 9 = 10	**b.** 3 + 7 = 10	**c.** 8 + 2 = 10
_____ + _____ = 11	_____ + _____ = 11	_____ + _____ = 11
d. 6 + 4 = 10	**e.** 5 + 5 = 10	**f.** 7 + 3 = 10
_____ + _____ = 11	_____ + _____ = 11	_____ + _____ = 11

4. Add. Tell which idea you use to add.

| Trick with nine |
| Trick with eight |
| "Just one more" than a sum with 10 |

a. 7 + 7 = _____   **b.** 9 + 7 = _____

c. 8 + 3 = _____   **d.** 6 + 7 = _____

e. 5 + 6 = _____   **f.** 5 + 8 = _____

g. 8 + 8 = _____   **h.** 4 + 9 = _____

| Doubles chart |
| Just one more than a double |
| I just know it! |

5. Solve.

a. Maria had $9. Then her mom gave her $5 for picking berries.
Then she bought ice cream for $2.
How much does Maria have now?

b. Ashley had 9 shirts and her brother Andy had 8. Then they both got
three new shirts from their aunt. Now, who has more shirts?
How many more?

c. Emily had $10. She bought colored pencils for $6 and a pretty
eraser for $1. Now how much money does she have?

d. Natalie and Eric went to play tennis. They had 8 tennis balls with them.
During the game they lost two balls, but they also found four more balls
near the tennis court that other people had lost.
Now how many tennis balls do they have?

6. Each time, *two* more is added than in the previous problem. Can you see the patterns?

a. 8 + 2 = _____	b. 5 + 3 = _____	c. 9 + 2 = _____	d. 7 + 3 = _____
8 + 4 = _____	5 + 5 = _____	9 + 4 = _____	7 + 5 = _____
8 + 6 = _____	5 + 7 = _____	9 + 6 = _____	7 + 7 = _____
8 + 8 = _____	5 + 9 = _____	9 + 8 = _____	7 + 9 = _____

7. Add and subtract. Start with the number in the bottom left-hand corner and follow the arrows.

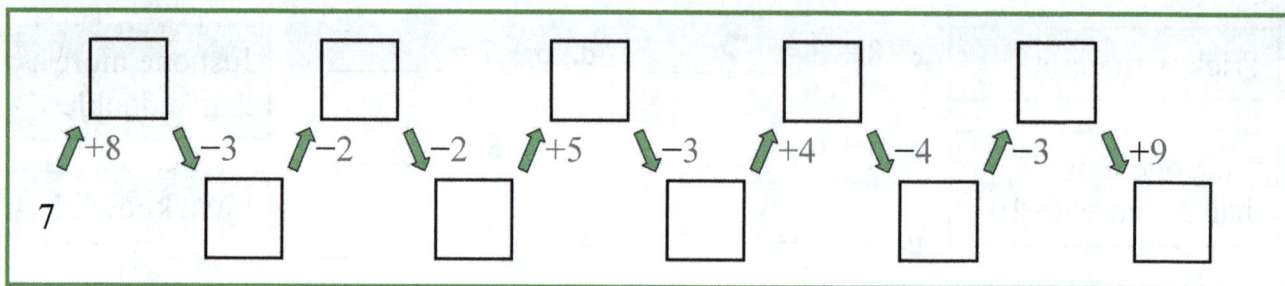

8. Count by tens.

a. 18, 28, _____, _____, _____, _____, _____, _____, _____

b. _____, _____, _____, _____, _____, 77, 87, _____, _____

9. What number goes in the triangle?

a. 6 + △ = 12	b. 8 + △ = 16	c. 6 + △ = 11
6 + △ = 13	9 + △ = 16	7 + △ = 11

10. Erica drew a line 9 cm long. Right after it she drew another, 6 cm long. How long are her two lines together? You draw them, too!

11. *A challenge!* Here are ALL the addition facts where the sum is more than 10. How many of them can you solve? We will study these more later in this book.

a.	b.	c.	d.
$8 + 8 =$ _____	$7 + 8 =$ _____	$7 + 7 =$ _____	$5 + 8 =$ _____
$2 + 9 =$ _____	$9 + 6 =$ _____	$9 + 8 =$ _____	$3 + 9 =$ _____
$7 + 5 =$ _____	$6 + 5 =$ _____	$7 + 4 =$ _____	$7 + 6 =$ _____
e.	f.	g.	h.
$9 + 4 =$ _____	$8 + 6 =$ _____	$9 + 2 =$ _____	$6 + 9 =$ _____
$4 + 8 =$ _____	$6 + 6 =$ _____	$8 + 5 =$ _____	$8 + 7 =$ _____
$6 + 7 =$ _____	$5 + 9 =$ _____	$5 + 7 =$ _____	$8 + 4 =$ _____
i.	j.	k.	l.
$9 + 3 =$ _____	$4 + 9 =$ _____	$9 + 9 =$ _____	$8 + 9 =$ _____
$4 + 7 =$ _____	$7 + 7 =$ _____	$6 + 8 =$ _____	$5 + 6 =$ _____
$9 + 5 =$ _____	$3 + 8 =$ _____	$6 + 6 =$ _____	$8 + 3 =$ _____

Puzzle Corner

What numbers can go into these puzzles?

	$+$		$= 13$
$+$		$+$	
	$+$		$= 11$
$=$		$=$	
15		9	

	$+$		$= 16$
$+$		$+$	
	$+$		$= 15$
$=$		$=$	
14		17	

17

Subtract to Ten

1. Subtract the "dots" that are not in the ten-group. You should only have ten left!

a.	b.	c.	d.
$12 - \underline{2} = 10$	$14 - \underline{\quad} = \boxed{10}$	$16 - \underline{\quad} = \underline{\quad}$	$15 - \underline{\quad} = \underline{\quad}$

2. Subtract the "ones" so that 10 is left.

a. $13 - \underline{\quad} = 10$ b. $17 - \underline{\quad} = \underline{\quad}$ c. $19 - \underline{\quad} = \underline{\quad}$

Subtracting in parts

Let's subtract $13 - 5$. First we subtract enough dots that we have only 10 left. So, first we take away 3 dots. $13 - 3 = 10$.

We still need to subtract 2 more. We subtract those from 10. There are 8 left.

$$13 - 5$$
$$/ \ \backslash$$
$$\boxed{13 - 3} - 2$$
$$= \underline{8}$$

---

3. First subtract enough dots so that you have only 10 left. Then subtract the rest.

a. $14 - 7$	b. $15 - 8$	c. $16 - 8$
$/ \ \backslash$	$/ \ \backslash$	$/ \ \backslash$
$\boxed{14 - 4} - 3$	$\boxed{15 - \underline{\quad}} - \underline{\quad}$	$\boxed{16 - \underline{\quad}} - \underline{\quad}$
$= \underline{\quad}$	$= \underline{\quad}$	$= \underline{\quad}$
d. $13 - 6$	e. $12 - 6$	f. $13 - 4$
$/ \ \backslash$	$/ \ \backslash$	$/ \ \backslash$
$\boxed{13 - \underline{\quad}} - \underline{\quad}$	$\boxed{12 - \underline{\quad}} - \underline{\quad}$	$\boxed{13 - \underline{\quad}} - \underline{\quad}$
$= \underline{\quad}$	$= \underline{\quad}$	$= \underline{\quad}$

4. First subtract to 10. Then subtract the rest.

a. $12 - 6$	b. $15 - 9$	c. $13 - 8$
$12 - \underline{\;2\;} - 4$	$15 - \underline{\quad} - \underline{\quad}$	$13 - \underline{\quad} - \underline{\quad}$
$= \underline{\quad}$	$= \underline{\quad}$	$= \underline{\quad}$
d. $13 - 7$	e. $14 - 7$	f. $12 - 4$
$13 - \underline{\quad} - \underline{\quad}$	$14 - \underline{\quad} - \underline{\quad}$	$12 - \underline{\quad} - \underline{\quad}$
$= \underline{\quad}$	$= \underline{\quad}$	$= \underline{\quad}$

5. First subtract those that are not in the ten-group.

a. $12 - 5 = \underline{\quad}$	b. $14 - 6 = \underline{\quad}$	c. $13 - 6 = \underline{\quad}$	d. $15 - 7 = \underline{\quad}$
e. $15 - 8 = \underline{\quad}$	f. $14 - 5 = \underline{\quad}$	g. $16 - 8 = \underline{\quad}$	h. $13 - 8 = \underline{\quad}$

6. Tom is 13, Juan is 8, and Alice is 9 years old.

   a. How many years older is Tom than Juan?

   b. How many years older is Tom than Alice?

   c. Two years later, how many years older is Tom now than Juan?

7. Finish this addition and subtraction "journey!"

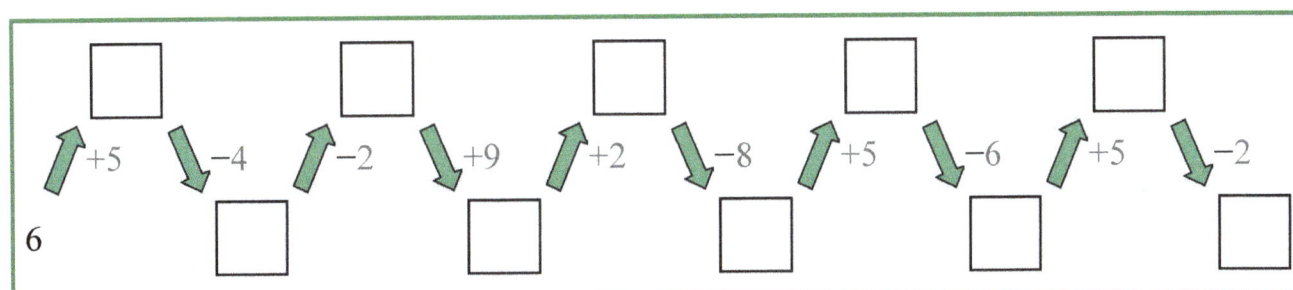

$$6 \quad \xrightarrow{+5} \quad \square \quad \xrightarrow{-4} \quad \square \quad \xrightarrow{-2} \quad \square \quad \xrightarrow{+9} \quad \square \quad \xrightarrow{+2} \quad \square \quad \xrightarrow{-8} \quad \square \quad \xrightarrow{+5} \quad \square \quad \xrightarrow{-6} \quad \square \quad \xrightarrow{+5} \quad \square \quad \xrightarrow{-2} \quad \square$$

Using Addition to Subtract

From the picture on the right we can write two additions and two subtractions: **a fact family**.

There are TWO parts that make up the total, 6 and 5. In addition, we add the parts and get the total.

In subtraction, we start with the total and take away one of the parts. What is left? The other part.

Fact family with 6, 5, and 11:

$6 + 5 = 11$ $\qquad$ $11 - 5 = 6$

$5 + 6 = 11$ $\qquad$ $11 - 6 = 5$

1. Write fact families.

a.

_____ + _____ = _____

_____ + _____ = _____

_____ − _____ = _____

_____ − _____ = _____

b.

_____ + _____ = _____

_____ + _____ = _____

_____ − _____ = _____

_____ − _____ = _____

2. For each addition, write two subtractions using the same numbers. Start with the TOTAL.

a. $8 + 4 =$ _____

_____ − _____ = _____

_____ − _____ = _____

b. $9 + 7 =$ _____

_____ − _____ = _____

_____ − _____ = _____

c. $7 + 6 =$ _____

_____ − _____ = _____

_____ − _____ = _____

3. For each subtraction write the matching addition.

a. $11 - 3 =$ _____

$3 +$ _____ $= 11$

b. $11 - 4 =$ _____

_____ $+$ _____ $= 11$

c. $12 - 3 =$ _____

_____ $+$ _____ $= 12$

When you have a subtraction problem like $15 - 9$ or $16 - 8$, try to find the matching addition.

$15 - 9 = ?$	**$16 - 8 = ?$**
Think: $9 + \underline{\quad} = 15$	Think: $8 + \underline{\quad} = 16$
9 and how many more is 15?	8 and what number makes 16?
Guess and check!	Guess and check!
Will $9 + 8$ work? Or $9 + 7$? Or $9 + 6$? Or $9 + 5$? You can use the trick with nine!	Will $8 + 5$ work? Or $8 + 6$? Or $8 + 7$?

4. Solve each subtraction by thinking about the matching addition.

a. $14 - 8 = \underline{\quad}$ $8 + \underline{\quad} = 14$	**b.** $15 - 7 = \underline{\quad}$ $7 + \underline{\quad} = 15$	**c.** $17 - 8 = \underline{\quad}$ $8 + \underline{\quad} = 17$
d. $12 - 8 = \underline{\quad}$ $\underline{\quad} + \underline{\quad} = 12$	**e.** $16 - 7 = \underline{\quad}$ $\underline{\quad} + \underline{\quad} = 16$	**f.** $13 - 7 = \underline{\quad}$ $\underline{\quad} + \underline{\quad} = 13$
g. $13 - 8 = \underline{\quad}$ $\underline{\quad} + \underline{\quad} = 13$	**h.** $11 - 7 = \underline{\quad}$ $\underline{\quad} + \underline{\quad} = 11$	**i.** $14 - 9 = \underline{\quad}$ $\underline{\quad} + \underline{\quad} = 14$

5. Doubles and doubles plus one more on the night sky! Solve. Also, find the matching additions and subtractions.

$8 + 8 = \underline{\qquad}$

$7 + 8 = \underline{\qquad}$

$15 - 7 = \underline{\qquad}$

$16 - 8 = \underline{\qquad}$

$14 - 7 = \underline{\qquad}$

$8 + 9 = \underline{\qquad}$

$17 - 8 = \underline{\qquad}$

$7 + 7 = \underline{\qquad}$

6. Now try to solve these subtractions by thinking of addition!

a. $12 - 8 =$ _____	b. $11 - 7 =$ _____	c. $13 - 9 =$ _____
d. $15 - 6 =$ _____	e. $18 - 9 =$ _____	f. $16 - 7 =$ _____

7. Solve.

a. Marsha had 15 crayons and Susana had 6. Marsha gave six of hers to Susana.
Now how many crayons does Marsha have?

And Susana?

Who has more?   How many more?

b. Judy counted seven stars in her drawing, and she thought, "That is not enough."
So, she drew eight more. How many stars are in her drawing now?

c. Matthew had $12. He bought a book for $7.
Now how much money does he have left?

d. John bought a toy truck for $6 and a toy backhoe for $8.
The shopkeeper said, "That makes $15."
John said, "That is not right, it makes $13."

Who is right?

8. Connect the problems to the right answer.

$14 - 9$		$13 - 9$		$12 - 6$		$14 - 6$
	7				9	
$12 - 5$		$14 - 7$		$15 - 7$		$16 - 7$
	5				6	
$12 - 8$		$11 - 6$		$18 - 9$		$15 - 9$
	4				8	
$13 - 8$		$15 - 8$		$16 - 7$		$13 - 7$

Review: Completing the Next Whole Ten

1. Write the previous and next **whole ten**. Then, circle the ten that is nearer the given number.

a. _____, 56, _____	b. _____, 72, _____	c. _____, 94, _____
d. _____, 37, _____	e. _____, 25, _____	f. _____, 31, _____

---

52 and how many more makes the next ten (60)? We can write $52 + \underline{\hspace{0.8cm}} = 60$.

You can solve it using a *helping* problem: **2 and how many more makes ten?**

The answer to both problems is the <u>same</u>. It is 8.

---

2. Complete the next ten. Below, write a helping problem using numbers within 0-10.

a. $17 + \underline{\hspace{0.8cm}} = 20$	b. $62 + \underline{\hspace{0.8cm}} = \underline{\hspace{1cm}}$	c. $94 + \underline{\hspace{0.8cm}} = \underline{\hspace{1cm}}$
$7 + \underline{\hspace{0.8cm}} = 10$	$2 + \underline{\hspace{0.8cm}} = \underline{\hspace{1cm}}$	$4 + \underline{\hspace{0.8cm}} = \underline{\hspace{1cm}}$

3. Complete the next ten. Think of the helping problem that uses numbers within 0-10.

a. $42 + \underline{\hspace{0.8cm}} = 50$	b. $34 + \underline{\hspace{0.8cm}} = \underline{\hspace{1cm}}$	c. $66 + \underline{\hspace{0.8cm}} = \underline{\hspace{1cm}}$
d. $61 + \underline{\hspace{0.8cm}} = \underline{\hspace{1cm}}$	e. $97 + \underline{\hspace{0.8cm}} = \underline{\hspace{1cm}}$	f. $83 + \underline{\hspace{0.8cm}} = \underline{\hspace{1cm}}$

4. Circle the even numbers.   8   9   12   15   10   19   11   6   17

5. Now pick the even numbers from the previous exercise, and write each of them as a double of some number.

a. _____ = _____ + _____	b. _____ = _____ + _____
c. _____ = _____ + _____	d. _____ = _____ + _____

6. Complete the next ten... and then go one more! Compare the two problems in each box.

a. 73 + _____ = 80	b. 35 + _____ = 40	c. 14 + _____ = 20
73 + _____ = 81	35 + _____ = 41	14 + _____ = 21

7. Find your way through the maze! Start at the top. You can only color a square if the sum is a whole ten (10, 20, 30, 40, 50, 60, 70, 80, 90, or 100).

13 + 6	54 + 6	73 + 8	45 + 7	99 + 4
15 + 9	14 + 8	15 + 5	13 + 6	32 + 7
45 + 7	73 + 7	64 + 5	82 + 9	16 + 7
30 + 12	39 + 1	74 + 6	73 + 9	52 + 7
46 + 7	32 + 7	31 + 9	86 + 4	65 + 4
92 + 4	21 + 8	24 + 7	22 + 8	32 + 6
83 + 6	11 + 7	98 + 2	57 + 3	17 + 9
44 + 9	12 + 8	95 + 6	38 + 5	53 + 9
71 + 9	34 + 4	36 + 7	19 + 4	28 + 11
53 + 7	29 + 2	26 + 6	78 + 6	32 + 5

8. Complete the next whole ten. These are more challenging.

a. 17 + _____ + 1 = 20	b. 35 + _____ + 2 = 40	c. 41 + _____ + 6 = 50
12 + _____ + 4 = 20	32 + _____ + 3 = 40	44 + _____ + 3 = 50
13 + _____ + 4 = 20	36 + _____ + 3 = 40	42 + _____ + 5 = 50

9. Find as many different sums as you can to make one hundred!

90 + _____ + _____ = 100	90 + _____ + _____ = 100	90 + _____ + _____ = 100
90 + _____ + _____ = 100	90 + _____ + _____ = 100	90 + _____ + _____ = 100
90 + _____ + _____ = 100	90 + _____ + _____ = 100	90 + _____ + _____ = 100

Review: Going Over Ten

Imagine that 8 wants to get some from 6 in order to make a ten. Six gives two to 8, and has only four left for itself!	Imagine that 9 wants to get some from 7 in order to make a ten. Seven gives one to 9, and has only six left for itself!

$8 + 6$

$8 + 2 + 4$

$10 + 4 = 14$

In the end, we have 10 and 4. We get 14.

$9 + 7$

$9 + 1 + 6$

$10 + 6 = 16$

In the end, we have 10 and 6. We get 16.

1. Circle all the blue balls and some of the red ones so that you get a ten. Then add the rest.

a. 8 + 4 $10 + \underline{\ 2\ } = \underline{\hspace{1.5cm}}$	**b.** 9 + 5 $10 + \underline{\hspace{1cm}} = \underline{\hspace{1.5cm}}$
c. 8 + 6 $10 + \underline{\hspace{1cm}} = \underline{\hspace{1.5cm}}$	**d.** 9 + 3 $10 + \underline{\hspace{1cm}} = \underline{\hspace{1.5cm}}$
e. 7 + 5 $10 + \underline{\hspace{1cm}} = \underline{\hspace{1.5cm}}$	**f.** 9 + 8 $10 + \underline{\hspace{1cm}} = \underline{\hspace{1.5cm}}$

2. Write a number on the empty line inside the balloon so that the numbers in the balloon make a ten. Then add the last number to 10.

a.	**b.**	**c.**
$7 + \underline{\ 3\ }$ $+ 2 = \underline{\hspace{1cm}}$	$5 + \underline{\hspace{1cm}}$ $+ 3 = \underline{\hspace{1cm}}$	$8 + \underline{\hspace{1cm}}$ $+ 4 = \underline{\hspace{1cm}}$
d.	**e.**	**f.**
$6 + \underline{\hspace{1cm}}$ $+ 4 = \underline{\hspace{1cm}}$	$9 + \underline{\hspace{1cm}}$ $+ 7 = \underline{\hspace{1cm}}$	$7 + \underline{\hspace{1cm}}$ $+ 5 = \underline{\hspace{1cm}}$

3. Fill in. Imagine that the first number wants to become a ten.

a. 8 + 7 / \ 8 + __2__ + __5__ 10 + __5__ = __15__	b. 8 + 9 / \ 8 + ____ + ____ 10 + ____ = ____	c. 8 + 5 / \ 8 + ____ + ____ 10 + ____ = ____
d. 9 + 4 / \ 9 + ____ + ____ 10 + ____ = ____	e. 9 + 6 / \ 9 + ____ + ____ 10 + ____ = ____	f. 9 + 9 / \ 9 + ____ + ____ 10 + ____ = ____

4. Add so that you get 10, 11, and 12. Notice the patterns!

a.	b.	c.	d.
8 + ____ = 10	7 + ____ = 10	9 + ____ = 10	6 + ____ = 10
8 + ____ = 11	7 + ____ = 11	9 + ____ = 11	6 + ____ = 11
8 + ____ = 12	7 + ____ = 12	9 + ____ = 12	6 + ____ = 12

5. Find the even numbers.

| 15   24   58   89   99 |
| 40   51   67   100   2 |

6. Solve the word problems. ALSO, write an addition & subtraction sentence for them!

a. You have $8 and you buy a toy for $5 and candy for $2.
How much money do you have now?

b. Lucy had $8. Then she found $5 in her piggy
bank, and her mom gave her $2.
How much money does she have now?

c. Matthew had $8. He spent $3 on a bottle of juice.
Later he found $2 in the street.
How much money does he have now?

Adding with 9

Imagine that 9 *really* wants to be a 10! It takes one from the other number (from 5). So, 9 becomes 10, and four dots are left over.

$$9 + 5 = 10 + 4 = 14$$

9 wants to be a 10! So, it takes one from the other number (from 3). So, 9 becomes 10, and two dots are left over.

$$9 + 3 = 10 + 2 = 12$$

Use the list on the right to practice. Don't write the answers there. Just point to different problems and say the answer aloud.

1. Circle the ten, then add..

a. $9 + 5$	b. $9 + 4$	c. $9 + 7$
$10 + 4 = $ _____	$10 + $ ____ $ = $ _____	$10 + $ ____ $ = $ _____
d. $9 + $ _____	e. $9 + $ _____	f. $9 + $ _____
$10 + $ ____ $ = $ _____	$10 + $ ____ $ = $ _____	$10 + $ ____ $ = $ _____

$9 + 1 = \square$
$9 + 2 = \square$
$9 + 3 = \square$
$9 + 4 = \square$
$9 + 5 = \square$
$9 + 6 = \square$
$9 + 7 = \square$
$9 + 8 = \square$
$9 + 9 = \square$

2. It is good to memorize the doubles, also. Fill in.

$2 + 2 = $ _____	$5 + 5 = $ _____	$8 + 8 = $ _____
$3 + 3 = $ _____	$6 + 6 = $ _____	$9 + 9 = $ _____
$4 + 4 = $ _____	$7 + 7 = $ _____	$10 + 10 = $ _____

3. Add to nine. Think how 9 wants to be a ten, and takes 1 from the other number.

a. $9 + 6$ $10 + 5 =$ _____	**b.** $9 + 8$ $10 +$ _____ $=$ _____	**c.** $9 + 5$ $10 +$ _____ $=$ _____
d. $9 + 7$ $10 +$ _____ $=$ _____	**e.** $9 + 9$ $10 +$ _____ $=$ _____	**f.** $9 + 3$ $10 +$ _____ $=$ _____

4. Practice the facts with nine. Do not write the answers down; just practice the sums.

$9 + 0 = \square$	$9 + 5 = \square$	$9 + 9 = \square$	$9 + 4 = \square$
$9 + 3 = \square$	$9 + 6 = \square$	$9 + 1 = \square$	
$9 + 7 = \square$	$9 + 8 = \square$	$9 + 2 = \square$	$9 + 10 = \square$

5. Add. Remember, you can add both ways. For example, $7 + 9$ is the same as $9 + 7$.

a. $9 + 4 =$ _____ $8 + 9 =$ _____ $9 + 5 =$ _____	**b.** $9 + 7 =$ _____ $4 + 9 =$ _____ $9 + 4 =$ _____	**c.** $3 + 9 =$ _____ $9 + 2 =$ _____ $9 + 9 =$ _____	**d.** $5 + 9 =$ _____ $8 + 9 =$ _____ $9 + 6 =$ _____

6. What is missing?

a. $9 + \boxed{} = 13$ $9 + \boxed{} = 15$	**b.** $9 + \boxed{} = 16$ $9 + \boxed{} = 14$	**c.** $\boxed{} + 9 = 17$ $\boxed{} + 9 = 11$

You can use this same "trick" with 19, 29, 39, 49, and so on. Imagine that 49 *really* wants to be 50, and so it "takes" 1 from the other number. Solve.

Puzzle Corner

a. $49 + 7 =$ _____   **b.** $59 + 5 =$ _____   **c.** $69 + 3 =$ _____

    $19 + 6 =$ _____ $89 + 9 =$ _____ $29 + 6 =$ _____

Adding with 8

Imagine that 8 wants to be a 10! It takes two from the other number (from 3). So, 8 becomes 10, and only 1 is left over.

$$8 + 3 = 10 + 1 = 11$$

8 wants to be a 10! So, it takes two from the other number (from 5). So, 8 becomes 10, and 3 are left over.

$$8 + 5 = 10 + 3 = 13$$

Use the list on the right to practice. Don't write the answers there. Just point to different problems and say the answer aloud.

1. Add. First, circle the ten.

a. $8 + 5$	**b.** $8 + 4$	**c.** $8 + \underline{\quad}$
$10 + 3 = \underline{\quad}$	$10 + \underline{\quad} = \underline{\quad}$	$10 + \underline{\quad} = \underline{\quad}$
d. $8 + \underline{\quad} =$	**e.** $8 + \underline{\quad} =$	**f.** $8 + \underline{\quad} =$
$10 + \underline{\quad} = \underline{\quad}$	$10 + \underline{\quad} = \underline{\quad}$	$10 + \underline{\quad} = \underline{\quad}$

$8 + 1 = \square$

$8 + 2 = \square$

$8 + 3 = \square$

$8 + 4 = \square$

$8 + 5 = \square$

$8 + 6 = \square$

$8 + 7 = \square$

$8 + 8 = \square$

$8 + 9 = \square$

2. It is good to memorize the doubles, also. Fill in.

$2 + 2 = \underline{\quad}$	$5 + 5 = \underline{\quad}$	$8 + 8 = \underline{\quad}$
$3 + 3 = \underline{\quad}$	$6 + 6 = \underline{\quad}$	$9 + 9 = \underline{\quad}$
$4 + 4 = \underline{\quad}$	$7 + 7 = \underline{\quad}$	$10 + 10 = \underline{\quad}$

Addition facts with eight. Do not write the answers down, but just practice the sums.

$8 + 0 =$ ☐	$8 + 5 =$ ☐	$8 + 8 =$ ☐	$8 + 9 =$ ☐
$8 + 3 =$ ☐	$8 + 7 =$ ☐	$8 + 1 =$ ☐	$8 + 4 =$ ☐
$8 + 10 =$ ☐	$8 + 1 =$ ☐	$8 + 6 =$ ☐	$8 + 2 =$ ☐

3. Add and fill in what is missing.

a. $8 + 4 =$ _____ $8 + 6 =$ _____	**b.** $7 + 8 =$ _____ $8 + 5 =$ _____	**c.** $3 + 8 =$ _____ $8 + 9 =$ _____
d. $8 +$ _____ $= 13$ $8 +$ _____ $= 15$	**e.** $8 +$ _____ $= 12$ $8 +$ _____ $= 16$	**f.** _____ $+ 8 = 11$ _____ $+ 8 = 14$

4. **a.** Jenny ate 8 strawberries, and Jack ate 5 more than what Jenny did.
   How many strawberries did Jack eat?

   b. Ashley is 13 years old, and Maryann is 5.
   How many years older is Ashley than Maryann?

5. Find the patterns and continue them.

a. $8 + 2 =$ _____ $8 + 4 =$ _____ $8 + 6 =$ _____ $8 +$ _____ $=$ _____ _____ $+$ _____ $=$ _____ _____ $+$ _____ $=$ _____ _____ $+$ _____ $=$ _____	**b.** $18 + 2 =$ _____ $18 + 4 =$ _____ $18 + 6 =$ _____ $18 +$ _____ $=$ _____ _____ $+$ _____ $=$ _____ _____ $+$ _____ $=$ _____ _____ $+$ _____ $=$ _____	**c.** $\frac{1}{2}$ of 0 is _____. $\frac{1}{2}$ of 2 is _____. $\frac{1}{2}$ of 4 is _____. $\frac{1}{2}$ of _____ is _____. $\frac{1}{2}$ of _____ is _____. $\frac{1}{2}$ of _____ is _____. $\frac{1}{2}$ of _____ is _____.

Adding with 7

We have already studied these facts:

$7 + 8 =$ _____ $8 + 7 =$ _____

$7 + 9 =$ _____ $9 + 7 =$ _____

$7 + 10 =$ _____ $10 + 7 =$ _____

These are the new facts with 7:

$7 + 4 =$ _____ $7 + 6 =$ _____

$7 + 5 =$ _____ $7 + 7 =$ _____

<u>Tricks for remembering facts with 7</u>

- $7 + 7 = 14$ is one of the doubles. Memorize all the doubles!
  But if you forget, you can do $5 + 5 = 10$, then $6 + 6 = 12$,
  and *then* $7 + 7 = 14$.

- $7 + 6$ is *just one more* than the doubles fact $6 + 6 = 12$. So, it
  is 13. Or, $7 + 6$ is *just one less* than the doubles fact $7 + 7 = 14$.

- $7 + 4$ is *just one more* than the ten-fact $7 + 3 = 10$. So, $7 + 4$ is 11.

- $7 + 5$ is just one more than $7 + 4$, or just one less than $7 + 6$,
  so if you remember those, you can figure out $7 + 5$, too.
  Or maybe you have your own trick for it!

$7 + 1 = \square$

$7 + 2 = \square$

$7 + 3 = \square$

$7 + 4 = \square$

$7 + 5 = \square$

$7 + 6 = \square$

$7 + 7 = \square$

$7 + 8 = \square$

$7 + 9 = \square$

Use the list on the right to practice. Don't write the answers there.
Just point to different problems and say the answer aloud.

1. Let's practice doubles—and doubles plus **one more**.
   Notice: the answer is also just one more!

a. $6 + 6 =$ _____	**b.** $7 + 7 =$ _____	**c.** $8 + 8 =$ _____
$6 + 7 =$ _____	$7 + 8 =$ _____	$8 + 9 =$ _____
d. $9 + 9 =$ _____	**e.** $5 + 5 =$ _____	**f.** $4 + 4 =$ _____
$9 + 10 =$ _____	$6 + 5 =$ _____	$4 + 5 =$ _____

Addition facts with seven. Do not write the answers down, but just practice the sums.

$7 + 0 = \square$	$7 + 5 = \square$	$7 + 6 = \square$	$7 + 9 = \square$
$7 + 3 = \square$	$7 + 9 = \square$	$7 + 7 = \square$	$7 + 4 = \square$
$7 + 10 = \square$	$7 + 8 = \square$	$7 + 1 = \square$	$7 + 2 = \square$

2. Fill in the missing numbers.

a. $7 + 4 = \underline{\hspace{1cm}}$	**b.** $8 + 7 = \underline{\hspace{1cm}}$	**c.** $7 + \underline{\hspace{1cm}} = 14$	**d.** $7 + \underline{\hspace{1cm}} = 12$
$6 + 7 = \underline{\hspace{1cm}}$	$7 + 10 = \underline{\hspace{1cm}}$	$7 + \underline{\hspace{1cm}} = 13$	$7 + \underline{\hspace{1cm}} = 16$
$7 + 5 = \underline{\hspace{1cm}}$	$3 + 7 = \underline{\hspace{1cm}}$	$7 + \underline{\hspace{1cm}} = 15$	$7 + \underline{\hspace{1cm}} = 11$
e. $7 + 7 = \underline{\hspace{1cm}}$	**f.** $4 + 7 = \underline{\hspace{1cm}}$	**g.** $8 + \underline{\hspace{1cm}} = 13$	**h.** $\underline{\hspace{1cm}} + 7 = 17$
$9 + 7 = \underline{\hspace{1cm}}$	$7 + 9 = \underline{\hspace{1cm}}$	$8 + \underline{\hspace{1cm}} = 16$	$\underline{\hspace{1cm}} + 7 = 10$
$7 + 8 = \underline{\hspace{1cm}}$	$3 + 7 = \underline{\hspace{1cm}}$	$8 + \underline{\hspace{1cm}} = 17$	$\underline{\hspace{1cm}} + 7 = 12$

3. Try these boxes!

Add 7 each time.
Add 8 each time.
Add 9 each time.

$+7$

4	11
7	___
8	___
10	___
5	___
9	___

$+8$

3	11
6	___
5	___
7	___
2	___
4	___

$+9$

2	___
4	___
7	___
8	___
3	___
5	___

Adding with 6

$6 + 5 =$ ___	$6 + 6 =$ ___
This is **just one more** than $5 + 5 = 10$.	This is one of the doubles!

Here are addition facts where we add to six. Do not write the answers down. Just go over the problems until you remember them easily.

$6 + 0 =$ ☐	$6 + 5 =$ ☐	$6 + 9 =$ ☐	$6 + 6 =$ ☐
$6 + 3 =$ ☐	$6 + 7 =$ ☐	$6 + 4 =$ ☐	$6 + 8 =$ ☐
$6 + 10 =$ ☐	$6 + 1 =$ ☐	$6 + 2 =$ ☐	

1. Fill in the missing numbers.

a.	b.	c.	d.
$6 + 4 =$ ___	$6 + 8 =$ ___	$6 +$ ___ $= 14$	___ $+ 6 = 12$
$6 + 6 =$ ___	$6 + 9 =$ ___	$6 +$ ___ $= 16$	___ $+ 6 = 15$
$6 + 5 =$ ___	$6 + 7 =$ ___	$6 +$ ___ $= 12$	___ $+ 6 = 11$

e.	f.	g.	h.
$5 + 6 =$ ___	$9 + 6 =$ ___	$7 +$ ___ $= 14$	___ $+ 6 = 13$
$6 + 7 =$ ___	$8 + 6 =$ ___	$8 +$ ___ $= 14$	___ $+ 6 = 14$
$4 + 6 =$ ___	$6 + 6 =$ ___	$9 +$ ___ $= 14$	___ $+ 6 = 15$

Trick! When you add three or four numbers, first add the numbers that make ten. It makes adding easier!

$$8 + 6 + 4 \qquad\qquad 5 + 3 + 2 + 5$$

$$= 8 + 10 = 18 \qquad = 10 + 5 = 15$$

2. Add. *First* find the numbers that make 10. You can circle or color them. Then add the rest. This is like hide-and-seek! Where are those numbers lurking that make ten?

a.	b.	c.
$1 + 6 + 9 = $ _____	$3 + 6 + 7 + 2 = $ _____	$6 + 5 + 1 + 4 = $ _____
$6 + 8 + 2 = $ _____	$1 + 5 + 5 + 7 = $ _____	$8 + 3 + 2 + 6 = $ _____
$5 + 7 + 5 = $ _____	$2 + 7 + 8 + 2 = $ _____	$9 + 6 + 1 + 4 = $ _____

3. Solve the word problems.

a. There were some apples on the table. Children came in and ate 5 apples. Later, mom saw 7 apples left on the table. How many apples had there been at first?

b. Jeremy had $12. He bought a toy truck, and then he had $6 left. How much did the toy truck cost?

c. Mom bought a bunch of bananas. She ate one, Dad ate two, and the children ate two. Then there were four bananas left. How many bananas did Mom buy?

d. Mike solved 9 math problems. Scott solved 5 more than Mike. How many did Scott solve?

e. Elena solved 14 math problems and Ashley solved 7. How many more did Elena solve than Ashley?

Review—Facts with 6, 7, and 8

1. Here are the 20 addition facts with single-digit numbers where the sum is between 10 and 20. Connect the problems to the right answer.

$6 + 6$

$5 + 8$ 11

$9 + 5$ 12

$5 + 6$ 13

$3 + 9$

$7 + 7$ 14

$8 + 3$

$8 + 6$

$5 + 7$ 15

$9 + 2$ 16

$4 + 7$

$9 + 4$ 17

$6 + 7$ 18

$4 + 8$

$9 + 9$

$7 + 9$

$8 + 7$

$9 + 8$

$8 + 8$

$6 + 9$

2. Figure out the pattern and continue it.

a.	b.	c.
$9 + \underline{} = 19$	$\underline{} + 16 = 17$	$6 + \underline{} = 6$
$8 + \underline{} = 18$	$\underline{} + 14 = 17$	$6 + \underline{} = 8$
$7 + \underline{} = 17$	$\underline{} + 12 = 17$	$6 + \underline{} = 10$
$\underline{} + \underline{} = \underline{}$	$\underline{} + \underline{} = \underline{}$	$\underline{} + \underline{} = \underline{}$
$\underline{} + \underline{} = \underline{}$	$\underline{} + \underline{} = \underline{}$	$\underline{} + \underline{} = \underline{}$
$\underline{} + \underline{} = \underline{}$	$\underline{} + \underline{} = \underline{}$	$\underline{} + \underline{} = \underline{}$
$\underline{} + \underline{} = \underline{}$	$\underline{} + \underline{} = \underline{}$	$\underline{} + \underline{} = \underline{}$
$\underline{} + \underline{} = \underline{}$	$\underline{} + \underline{} = \underline{}$	$\underline{} + \underline{} = \underline{}$

3. Fill in the addition table.

+	6	8	4	5	7	3	9
7							
9							
5							

4. Solve.

a. A herd of elephants was feeding on the grass. Four of them left, but fourteen stayed feeding. How many elephants are in the herd?

b. Sarah has five more dolls than Annie. Sarah has 10 dolls. How many does Annie have?

Hint 1: Draw Sarah's dolls. Hint 2: Think which girl has more dolls. Should you draw more or fewer dolls for Annie?

c. Ronnie and Luis emptied waste baskets. Ronnie emptied four more waste baskets than Luis. Luis emptied five baskets. How many did Ronnie empty?

Hint 1: Draw Luis's baskets. Hint 2: Think which boy emptied more of them. Should you draw more or fewer baskets for Ronnie?

d. Cynthia ate 10 peanuts. Marie ate 7 more than Cynthia. How many did Marie eat?

5. Add. In some problems, you can find numbers that *make a ten*.

a.	b.	c.
$6 + 6 + 2 = $ _____	$8 + 6 + 3 = $ _____	$6 + 2 + 3 + 7 = $ _____
$1 + 4 + 9 = $ _____	$2 + 2 + 8 = $ _____	$3 + 6 + 7 + 2 = $ _____

Difference and How Many More

The difference or distance between two numbers means <u>how far apart</u> they are from each other on the number line. The difference between 3 and 12 is 9, because they are NINE steps apart.

The difference is 9.

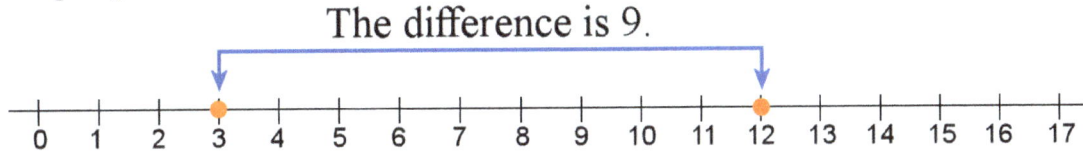

1. Find the differences between these numbers using the number line above.

   a. difference between 10 and 6: _____   **b.** difference between 12 and 8: _____

   c. difference between 14 and 2: _____   **d.** difference between 17 and 6: _____

We can solve the difference between two numbers by **subtracting**.

What is the difference between 10 and 4? Subtract 10 − 4 = 6. The difference is 6.

2. Write a subtraction to find the difference between the numbers.

a. The difference between 10 and 4	**b.** The difference between 2 and 9	**c.** The difference between 8 and 3
_____ − _____ = _____	_____ − _____ = _____	_____ − _____ = _____
d. The difference between 20 and 50	**e.** The difference between 10 and 90	**f.** The difference between 19 and 8
_____ − _____ = _____	_____ − _____ = _____	_____ − _____ = _____

3. Solve the subtractions by thinking of the <u>distance between the numbers</u>—how far apart they are from each other.

a.	**b.**	**c.**	**d.**
20 − 16 = _____	40 − 38 = _____	65 − 61 = _____	36 − 31 = _____
e.	**f.**	**g.**	**h.**
100 − 99 = _____	87 − 84 = _____	55 − 50 = _____	79 − 78 = _____

You can also solve the difference between two numbers by thinking of addition: how many more do you need to add to the one number to get the other?

For example, to find the difference between 12 and 7, think: $7 + \underline{\quad} = 12$. ("7 and how many more makes 12?") The answer is 5.

4. Write a "*how many more*" addition to find the difference between the numbers.

a. The difference between 10 and 6	**b.** The difference between 6 and 12
$6 + \underline{\quad} = 10$	$6 + \underline{\quad} = 12$
c. The difference between 15 and 8	**d.** The difference between 4 and 11
$\underline{\quad} + \underline{\quad} = \underline{\quad}$	$\underline{\quad} + \underline{\quad} = \underline{\quad}$

5. Subtract. Think how far apart the two numbers are from each other.

$+3$	$+ \underline{\quad}$	$+ \underline{\quad}$
a. $15 - 12 = \underline{\quad}$	**b.** $11 - 9 = \underline{\quad}$	**c.** $16 - 11 = \underline{\quad}$
12 and *how many more* makes 15?	9 and *how many more* makes 11?	11 and *how many more* makes 16?

There are two ways to find a difference between two numbers:

(1) Subtraction	**(2) A "*how many more*" addition**
Find the difference between 100 and 2. It is easier to subtract $100 - 2 = 98$. The difference is 98.	Find the difference between 100 and 95. It is easier to think: $95 + \underline{\quad} = 100$. The difference is 5.

6. Find the differences.

a. The difference between 60 and 56	**b.** The difference between 22 and 20
c. The difference between 35 and 1	**d.** The difference between 67 and 3
e. The difference between 50 and 30	**f.** The difference between 40 and 100

Whenever a word problem asks "*how many more*," you can solve it in two ways.
You can either subtract, or you can write a "*how many more*" addition.
Either way, you are finding the difference between the two numbers.

7. Solve the word problems.

a. Jane is on page 20 and Boyd is on page 17 of the same book.
How many more pages has Jane read?

b. Mom has one dozen eggs plus five in another carton. A dozen means 12.
How many eggs does Mom have?

c. Barb is reading a 50-page book. She is on page 42.
How many more pages does she have left to read?

d. Janet worked in the garden for 2 hours in the morning and 3 hours
in the afternoon. Andy worked for 8 hours in the shop.
Who worked more hours?

How many more?

e. Betty is going batty with flies! She killed 28 flies. Her husband killed 5 flies.
How many more did she kill than him?

f. The next day, Betty was again going batty with flies. She killed 5 flies
in the living room, 12 in the kitchen, and 2 in her room.
How many flies did she kill in total?

g. Matthew had $12 and Bob had $6. Then both brothers worked helping Dad
in the garden. Matthew earned $5 and Bob earned $9.
Now, who has more money?

How much more?

Number Rainbows—11 and 12

This is a number rainbow for 11. If two numbers are connected with an arc, they add up to 11. Use the number rainbow to help you with addition and subtraction facts!

1. Practice subtraction from 11. Don't write the answers; just think them in your head.

11

$$0 \quad 1 \quad 2 \quad 3 \quad 4 \quad 5 \quad 6 \quad 7 \quad 8 \quad 9 \quad 10 \quad 11$$

$11 - 6 = \square$	$11 - 7 = \square$	$11 - 8 = \square$	$11 - 2 = \square$
$11 - 3 = \square$	$11 - 9 = \square$	$11 - 4 = \square$	$11 - 5 = \square$

2. Similarly, practice subtraction from 12.

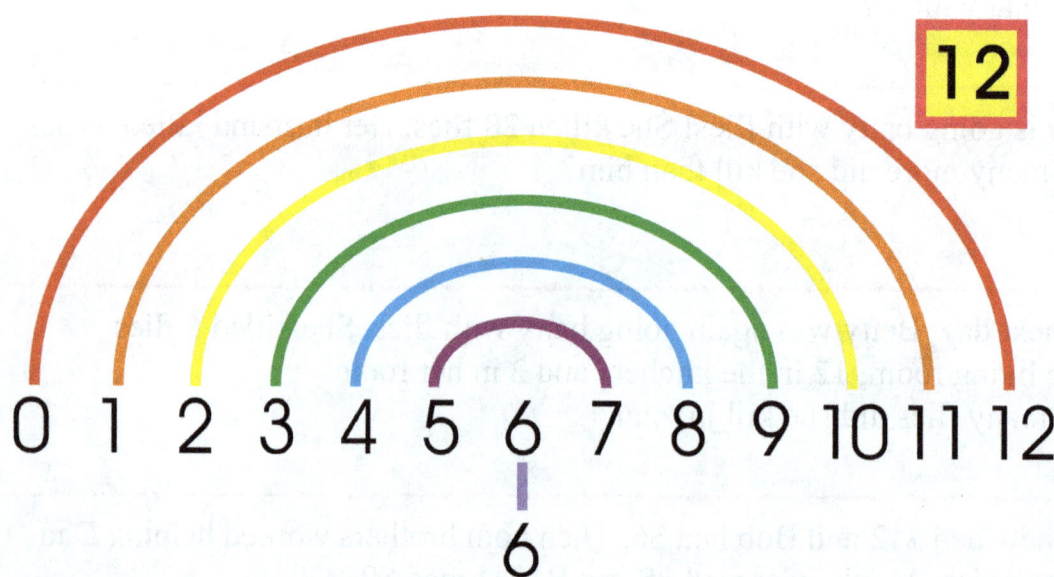

12

$$0 \quad 1 \quad 2 \quad 3 \quad 4 \quad 5 \quad 6 \quad 7 \quad 8 \quad 9 \quad 10 \quad 11 \quad 12$$

6

$12 - 5 = \square$	$12 - 7 = \square$	$12 - 10 = \square$	$12 - 6 = \square$
$12 - 9 = \square$	$12 - 4 = \square$	$12 - 3 = \square$	$12 - 8 = \square$

3. Fill and color the number rainbows. Don't look at the previous page!
   Then practice the subtraction problems.

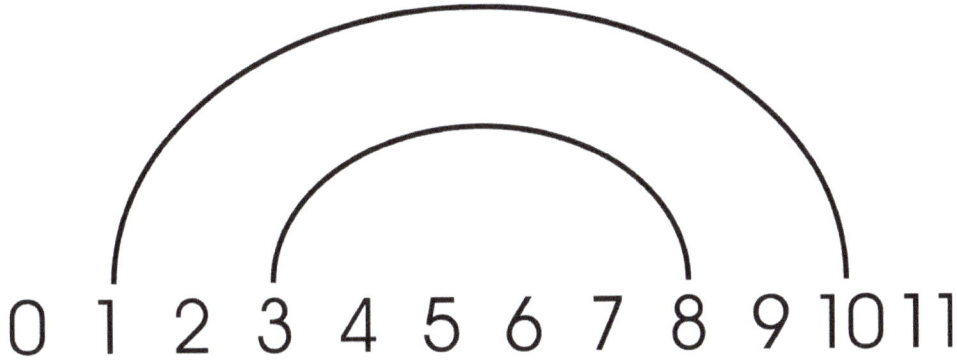

0 1 2 3 4 5 6 7 8 9 10 11

11 − 4 = □   11 − 2 = □   11 − 3 = □   11 − 9 = □

11 − 8 = □   11 − 5 = □   11 − 6 = □   11 − 7 = □

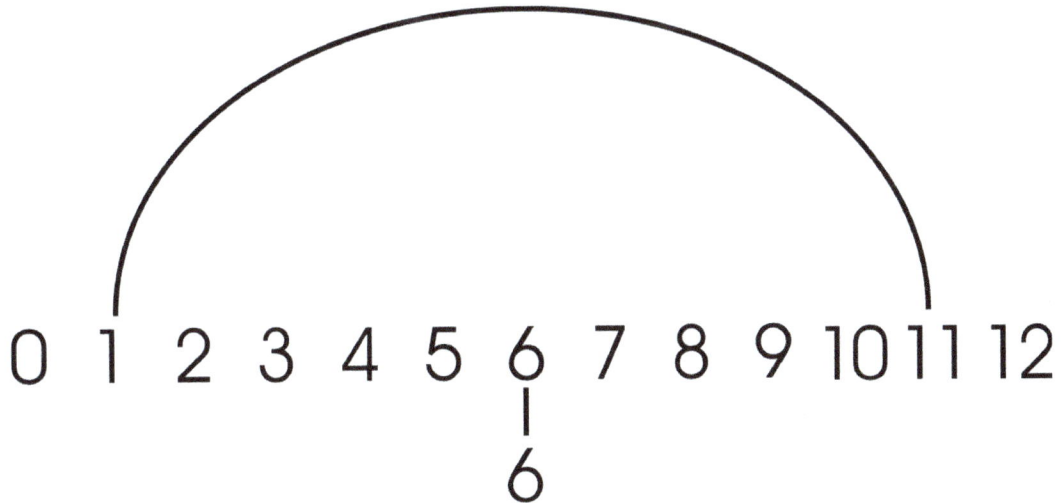

0 1 2 3 4 5 6 7 8 9 10 11 12

|
6

12 − 8 = □   12 − 3 = □   12 − 4 = □   12 − 9 = □

12 − 6 = □   12 − 10 = □   12 − 7 = □   12 − 5 = □

For more practice, make your own number rainbows and subtractions on blank paper!

Fact Families with 11

1. Fill in. In each fact family, color enough marbles to equal the first number. Then use another color to color the rest.

Fact families with 11		
10, 1, and 11	$10 + 1 =$ _____ $1 + 10 =$ _____	$11 - 10 =$ _____ $11 - 1 =$ _____
9, _____, and 11	$9 +$ _____ $= 11$ _____ $+$ _____ $=$ _____	_____ $-$ _____ $=$ _____ _____ $-$ _____ $=$ _____
8, _____, and 11	$8 +$ _____ $= 11$ _____ $+$ _____ $=$ _____	_____ $-$ _____ $=$ _____ _____ $-$ _____ $=$ _____
7, _____, and 11	$7 +$ _____ $= 11$ _____ $+$ _____ $=$ _____	_____ $-$ _____ $=$ _____ _____ $-$ _____ $=$ _____
6, _____, and 11	$6 +$ _____ $= 11$ _____ $+$ _____ $=$ _____	_____ $-$ _____ $=$ _____ _____ $-$ _____ $=$ _____

2. Check yourself! Can you subtract quickly without looking above?

a. $11 - 10 =$ _____ $11 - 9 =$ _____ $11 - 6 =$ _____ $11 - 8 =$ _____	**b.** $11 - 2 =$ _____ $11 - 4 =$ _____ $11 - 5 =$ _____ $11 - 7 =$ _____	**c.** $11 - 3 =$ _____ $11 - 6 =$ _____ $11 - 9 =$ _____ $11 - 4 =$ _____

Fact Families with 12

1. Fill in. In each fact family, color enough marbles to equal the first number. Then use another color to color the rest.

Fact families with 12		
10, 2, and 12 ⚾⚾⚾⚾⚾⚾⚾⚾⚾⚾ ⚾⚾	$10 + 2 =$ ____ $2 + 10 =$ ____	$12 - 10 =$ ____ $12 - 2 =$ ____
9, ____, and 12 ⚾⚾⚾⚾⚾⚾⚾⚾⚾ ⚾⚾	$9 +$ ____ $= 12$ ____ $+$ ____ $=$ ____	____ $-$ ____ $=$ ____ ____ $-$ ____ $=$ ____
8, ____, and 12 ⊗⊗○○○○○○○○ ⊗⊗	$8 +$ ____ $= 12$ ____ $+$ ____ $=$ ____	____ $-$ ____ $=$ ____ ____ $-$ ____ $=$ ____
7, ____, and 12 ⊗⊗⊗○○○○○○○ ⊗⊗	____ $+$ ____ $=$ ____ ____ $+$ ____ $=$ ____	____ $-$ ____ $=$ ____ ____ $-$ ____ $=$ ____
6, ____, and 12 ⊗⊗⊗○○○○○○○ ⊗⊗	____ $+$ ____ $=$ ____	____ $-$ ____ $=$ ____

2. Check yourself! Can you subtract quickly from 12 and from 11 without looking above?

a.	b.	c.	d.
$12 - 4 =$ ____	$11 - 8 =$ ____	$12 - 6 =$ ____	$12 - 3 =$ ____
$11 - 9 =$ ____	$12 - 7 =$ ____	$11 - 4 =$ ____	$12 - 10 =$ ____
$12 - 8 =$ ____	$11 - 3 =$ ____	$12 - 9 =$ ____	$11 - 5 =$ ____
$11 - 6 =$ ____	$12 - 5 =$ ____	$12 - 4 =$ ____	$11 - 7 =$ ____

3. Let's practice "how many more" additions! Remember the fact families with 11 and 12.

a. $6 + \underline{\hspace{1cm}} = 11$	**b.** $7 + \underline{\hspace{1cm}} = 12$	**c.** $\underline{\hspace{1cm}} + 9 = 11$	**d.** $\underline{\hspace{1cm}} + 6 = 12$
$8 + \underline{\hspace{1cm}} = 11$	$8 + \underline{\hspace{1cm}} = 12$	$\underline{\hspace{1cm}} + 7 = 11$	$\underline{\hspace{1cm}} + 9 = 12$

4. *Explain* how you can use *addition* to solve a subtraction problem, such as $11 - 8$.

5. Find the pattern and continue it.

a. $16 - 1 = \underline{\hspace{1cm}}$	**b.** $0 + 17 = \underline{\hspace{1cm}}$	**c.** $15 - 1 = \underline{\hspace{1cm}}$
$16 - 3 = \underline{\hspace{1cm}}$	$2 + 15 = \underline{\hspace{1cm}}$	$15 - 3 = \underline{\hspace{1cm}}$
$16 - 5 = \underline{\hspace{1cm}}$	$4 + 13 = \underline{\hspace{1cm}}$	$15 - 5 = \underline{\hspace{1cm}}$
$\underline{\hspace{1cm}} - \underline{\hspace{1cm}} = \underline{\hspace{1cm}}$	$\underline{\hspace{1cm}} + \underline{\hspace{1cm}} = \underline{\hspace{1cm}}$	$\underline{\hspace{1cm}} - \underline{\hspace{1cm}} = \underline{\hspace{1cm}}$
$\underline{\hspace{1cm}} - \underline{\hspace{1cm}} = \underline{\hspace{1cm}}$	$\underline{\hspace{1cm}} + \underline{\hspace{1cm}} = \underline{\hspace{1cm}}$	$\underline{\hspace{1cm}} - \underline{\hspace{1cm}} = \underline{\hspace{1cm}}$
$\underline{\hspace{1cm}} - \underline{\hspace{1cm}} = \underline{\hspace{1cm}}$	$\underline{\hspace{1cm}} + \underline{\hspace{1cm}} = \underline{\hspace{1cm}}$	$\underline{\hspace{1cm}} - \underline{\hspace{1cm}} = \underline{\hspace{1cm}}$
$\underline{\hspace{1cm}} - \underline{\hspace{1cm}} = \underline{\hspace{1cm}}$	$\underline{\hspace{1cm}} + \underline{\hspace{1cm}} = \underline{\hspace{1cm}}$	$\underline{\hspace{1cm}} - \underline{\hspace{1cm}} = \underline{\hspace{1cm}}$

Puzzle Corner   Imagine 14 baby blocks in three stacks. One stack has 6 and the third stack has 4. How many are in the middle stack?

We can write an addition where one number is missing: $6 + \underline{\hspace{1cm}} + 4 = 14$.
Figure out a way to solve this problem! Then solve the rest of the problems below.

a. $6 + \underline{\hspace{1cm}} + 4 = 14$	**b.** $2 + \underline{\hspace{1cm}} + 2 = 8$	**c.** $10 + \underline{\hspace{1cm}} + 4 = 17$
$8 + \underline{\hspace{1cm}} + 3 = 13$	$3 + \underline{\hspace{1cm}} + 3 = 9$	$10 + \underline{\hspace{1cm}} + 2 = 15$

Number Rainbows—13 and 14

1. Fill in and color the number rainbows. Then practice the subtractions.

13

0 1 2 3 4 5 6 7 8 9 10 11 12 13

13 − 7 = ☐   13 − 4 = ☐   13 − 9 = ☐   13 − 10 = ☐

13 − 5 = ☐   13 − 6 = ☐   13 − 11 = ☐   13 − 8 = ☐

14

0 1 2 3 4 5 6 7 8 9 10 11 12 13 14

7

14 − 8 = ☐   14 − 3 = ☐   14 − 7 = ☐   14 − 6 = ☐

14 − 5 = ☐   14 − 9 = ☐   14 − 11 = ☐   14 − 4 = ☐

For more practice, make your own number rainbows and subtractions on blank paper!

Fact Families - 13 and 14

1. Fill in. In each fact family, color the marbles so they match the numbers in it.

Fact families with 13		
10, 3, and 13 ●●●●●●●●●● ●●●	$10 + 3 =$ ____ $3 + 10 =$ ____	$13 - 10 =$ ____ $13 - 3 \ =$ ____
9, _____, and 13 ●●●●●●●●● ●●●	$9 +$ ____ $= 13$ ____ $+$ ____ $=$ ____	____ $-$ ____ $=$ ____ ____ $-$ ____ $=$ ____
8, ____, and 13 ◉◉◉◉○○○○○○ ◉◉◉	$8 +$ ____ $= 13$ ____ $+$ ____ $=$ ____	____ $-$ ____ $=$ ____ ____ $-$ ____ $=$ ____
7, ____, and 13 ◉◉◉○○○○○○○ ◉◉◉	$7 +$ ____ $= 13$ ____ $+$ ____ $=$ ____	____ $-$ ____ $=$ ____ ____ $-$ ____ $=$ ____

2. Connect with a line the problems that are from the same fact family. You don't need to write the answers.

$13 - 7 = $ ▪		$11 - 4 = $ ▪		$12 - 7 = $ ▪
$5 + $ ▪ $ = 12$		$11 - 8 = $ ▪		$13 - 6 = $ ▪
$11 - 3 = $ ▪		$5 + $ ▪ $ = 13$		$3 + $ ▪ $ = 12$
$8 + $ ▪ $ = 13$		$12 - 5 = $ ▪		$13 - 5 = $ ▪
$12 - 3 = $ ▪		$6 + $ ▪ $ = 13$		$3 + $ ▪ $ = 11$
$7 + $ ▪ $ = 11$		$9 + $ ▪ $ = 12$		$4 + $ ▪ $ = 11$

3. Fill in. In each fact family, color the marbles so they match the numbers in it.

Fact families with 14		
10, 4, and 14 ⚪⚪⚪⚪⚪⚪⚪⚪⚪⚪ ⚪⚪⚪⚪	10 + 4 = _____ 4 + 10 = _____	14 − 10 = _____ 14 − 4 = _____
9, _____, and 14 ⚪⚪⚪⚪⚪⚪⚪⚪⚪ ⚪⚪⚪⚪	9 + _____ = 14 _____ + _____ = _____	_____ − _____ = _____ _____ − _____ = _____
8, _____, and 14 ◯◯◯◯◯◯◯◯◯◯ ◯◯◯◯	8 + _____ = 14 _____ + _____ = _____	_____ − _____ = _____ _____ − _____ = _____
7, _____, and 14 ◯◯◯◯◯◯◯◯◯◯ ◯◯◯◯	7 + _____ = 14 _____ + _____ = _____	_____ − _____ = _____ _____ − _____ = _____

4. Subtract.

a. $13 - 8 =$ _____ $14 - 6 =$ _____	**b.** $13 - 5 =$ _____ $13 - 4 =$ _____	**c.** $12 - 7 =$ _____ $13 - 7 =$ _____	**d.** $12 - 9 =$ _____ $14 - 9 =$ _____

5. Find the missing numbers.

a. $9 + \boxed{} = 14$	**b.** $6 + \boxed{} = 14$	**c.** $6 + \boxed{} = 12$
d. $\boxed{} - 9 = 4$	**e.** $\boxed{} - 7 = 7$	**f.** $\boxed{} - 9 = 3$
g. $14 - \boxed{} = 8$	**h.** $12 - \boxed{} = 7$	**i.** $13 - \boxed{} = 8$

6. Solve the word problems.

a. Ted arranged his toy cars in rows. The first row had
seven cars, the second had seven, and the third row
had four. How many cars does Ted have?

b. If you have 14 strawberries and I have eight,
how many more do you have?

c. Dad has six cherries and Mom has five more than him.
How many cherries does Mom have?

d. At first Mom had 20 apples to make a pie,
but she gave each of the four children one apple
before she made the pie. How many apples
did she have left for the pie?

7. Figure out the patterns and continue them!

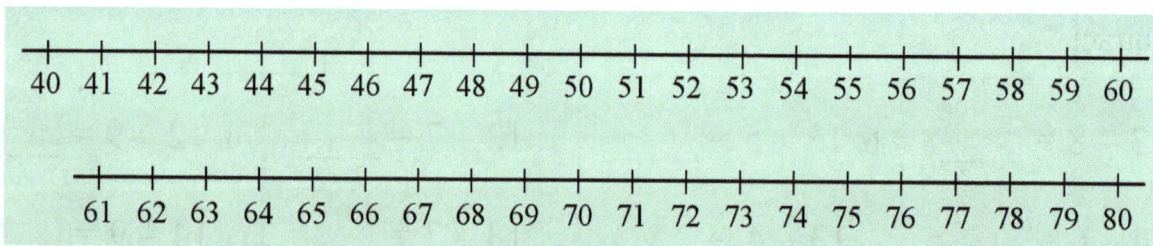

40 41 42 43 44 45 46 47 48 49 50 51 52 53 54 55 56 57 58 59 60

61 62 63 64 65 66 67 68 69 70 71 72 73 74 75 76 77 78 79 80

a.
40   48   56   64   72   ____ ____ ____ ____

b.
17   21   25   29   ____ ____ ____ ____

48

Fact Families with 15

1. Fill in. In each fact family, color the marbles so they match the numbers in it.

Fact families with 15		
10, 5, and 15	10 + 5 = _____	15 – 10 = _____
	5 + 10 = _____	15 – 5 = _____
9, _____, and 15	9 + _____ = 15	_____ – _____ = _____
	_____ + _____ = _____	_____ – _____ = _____
8, _____, and 15	8 + _____ = 15	_____ – _____ = _____
	_____ + _____ = _____	_____ – _____ = _____

2. Subtract.

a. 15 – 5 = _____	b. 15 – 8 = _____	c. 15 – 4 = _____
d. 15 – 9 = _____	e. 15 – 6 = _____	f. 15 – 7 = _____

3. Alice does not remember the answer to 15 – 9.
   Explain how she can solve it using *addition*.

4. Count by threes.

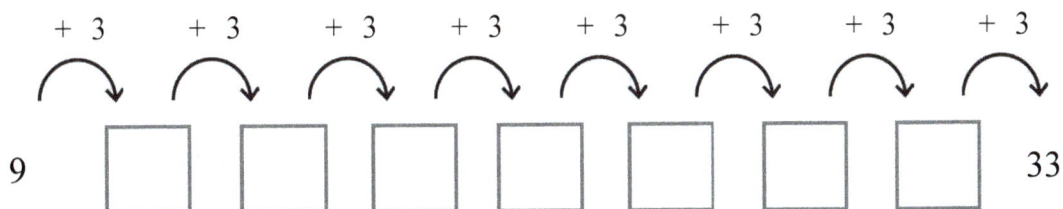

9   +3   +3   +3   +3   +3   +3   +3   +3   33

5. These word problems all have to do with "more." Draw a picture of how many things the one person has in the problem. Then think carefully **who** has more. Will you need to draw *more* or *fewer* things for the other person in the problem?

a. Michelle has 7 peaches and Jacob has three more than her. How many does Jacob have?

b. William has three more books than Ethan. William has 11 books. How many does Ethan have?

c. Noah picked 15 pine cones and Aiden picked 9. How many more did Noah pick than Aiden?

d. Emma picked 5 more pine cones than Sophia. If Emma picked 15, how many did Sophia pick?

6. Write each number as a double of some other number.

a. 6 = ____ + ____	**b.** 12 = ____ + ____	**c.** 10 = ____ + ____
d. 18 = ____ + ____	**e.** 20 = ____ + ____	**f.** 8 = ____ + ____

7. Judy picked 7 tomatoes from the garden, and John picked 9. Then they gave half of their tomatoes to a neighbor. How many did they keep?

8. Write or say all the even numbers from 0 to 20.

9. Find how much the things cost together.

a. bike, $28, and kite, $30	**b.** jeans, $47, and shoes, $30, and toy, $10
together $ _____	together $ _____

Fact Families with 16

1. Fill in. Color the marbles, using two colors, so that the coloring matches the numbers.

Fact families with 16		
10, 6, and 16	$10 + 6 =$ _____	$16 - 10 =$ _____
	$6 + 10 =$ _____	$16 - 6 =$ _____
9, _____, and 16	$9 + $ ____ $= 16$	____ $-$ ____ $=$ ____
	____ $+$ ____ $=$ _____	_____ $-$ ____ $=$ ____
8, ____, and 16	$8 + $ ____ $= 16$	_____ $-$ ____ $=$ ____
	____ $+$ ____ $=$ _____	_____ $-$ ____ $=$ ____

2. Subtract.

a.	b.	c.	d.
$15 - 10 =$ _____	$13 - 9 =$ _____	$14 - 8 =$ _____	$15 - 7 =$ _____
$13 - 10 =$ _____	$16 - 9 =$ _____	$13 - 8 =$ _____	$16 - 7 =$ _____
$16 - 10 =$ _____	$14 - 9 =$ _____	$16 - 8 =$ _____	$13 - 7 =$ _____

3. Connect the problems to the answer with a line.

$15 - 9$	3	$17 - 10$	7	$17 - 9$	
$14 - 9$	4	$16 - 9$	8	$16 - 6$	
$14 - 10$	5	$16 - 10$	9	$18 - 10$	
$13 - 9$	6	$18 - 9$	10	$19 - 9$	

51

4. Figure out the patterns and continue them!

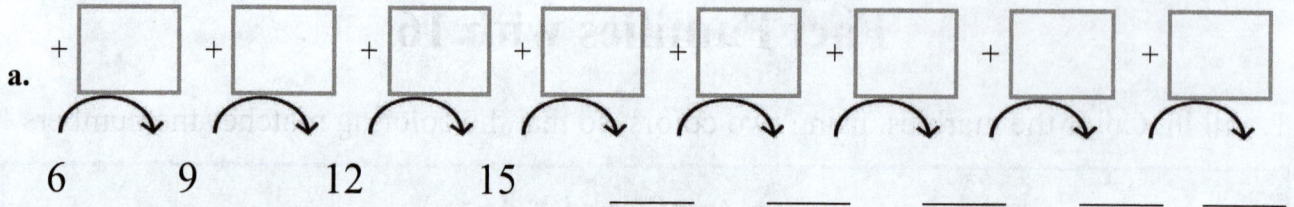

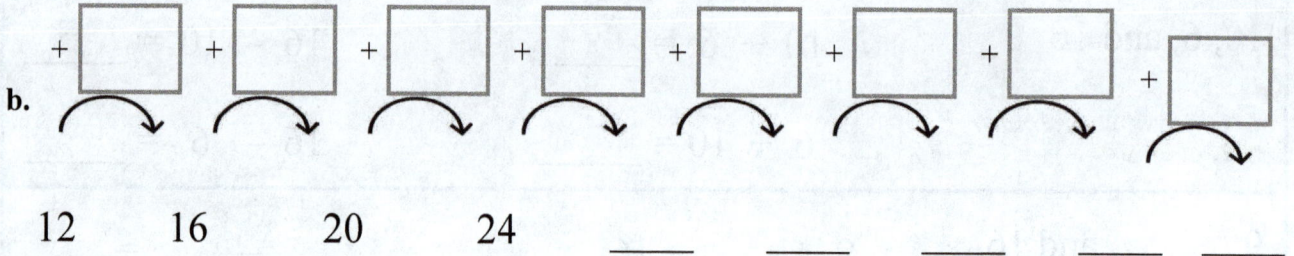

a.

$+\boxed{}$ $+\boxed{}$ $+\boxed{}$ $+\boxed{}$ $+\boxed{}$ $+\boxed{}$ $+\boxed{}$ $+\boxed{}$

6   9   12   15   ____   ____   ____   ____   ____

b.

$+\boxed{}$ $+\boxed{}$ $+\boxed{}$ $+\boxed{}$ $+\boxed{}$ $+\boxed{}$ $+\boxed{}$ $+\boxed{}$

12   16   20   24   ____   ____   ____   ____   ____

5. Solve the word problems.

a. A class has 24 children. Two of them were sick one day and two had to leave to go to the dentist. How many children were in class that day?

b. If you have $10, and Mom gives you $4 more, can you buy a book for $13?

c. You had $20 and you bought sandals for $17. How many dollars do you have left?

d. Erika has saved $12. She wants to buy a gift that costs $16. How much more money does she need?

e. Five boys came to play ball. Then, seven girls came. Then, one girl had to go home. Are there now more boys or girls playing ball?

How many more?

6. Compare and write < , > , or = .

a. $35\ \boxed{}\ 20 + 5$ **b.** $23 + 5\ \boxed{}\ 23 + 6$ **c.** $16 - 8\ \boxed{}\ 15 - 8$

d. $15\ \boxed{}\ 6 + 7$ **e.** $31 + 4\ \boxed{}\ 31 + 3$ **f.** $15 - 9\ \boxed{}\ 16 - 9$

Fact Families - 17 and 18

1. Fill in. Color the marbles, using two colors, so that the coloring matches the numbers.

Fact families with 17		
10, 7, and 17 	10 + 7 = _____ ____ + ____ = _____	17 − 10 = _____ _____ − ____ = _____
9, _____, and 17 	9 + ____ = 17 ____ + ____ = _____	_____ − ____ = _____ _____ − ____ = _____

Fact families with 18		
10, 8, and 18 	10 + 8 = _____ ____ + ____ = _____	18 − 10 = _____ _____ − ____ = _____
9, _____, and 18 	9 + ____ = 18 ____ + ____ = _____	_____ − ____ = _____ _____ − ____ = _____

2. Subtract, practicing the basic facts. Remember to think of fact families.

a.	b.	c.	d.
17 − 10 = _____	15 − 9 = _____	14 − 6 = _____	12 − 9 = _____
17 − 9 = _____	15 − 8 = _____	14 − 7 = _____	12 − 8 = _____
18 − 10 = _____	16 − 9 = _____	13 − 6 = _____	11 − 9 = _____
18 − 9 = _____	16 − 8 = _____	13 − 7 = _____	11 − 8 = _____

3. Write < , > , or = . Can you compare these without calculating?

a. $45 + 8$ ☐ $45 + 5$ b. $50 - 6$ ☐ $50 - 8$ c. $\frac{1}{2}$ of 12 ☐ 12

d. $\frac{1}{2}$ of 16 ☐ $\frac{1}{2}$ of 14   e. $27 - 6$ ☐ $27 - 3$ f. $\frac{1}{2}$ of 20 ☐ 10

4. Fill in the missing numbers.

a. $14 - 8 = \bigcirc$	b. $16 - 8 = \bigcirc$	c. $17 - 8 = \bigcirc$
d. $\bigcirc - 9 = 6$	e. $\bigcirc - 8 = 7$	f. $\bigcirc - 4 = 8$
g. $17 - \bigcirc = 9$	h. $18 - \bigcirc = 9$	i. $15 - \bigcirc = 6$

5. Solve the word problems.

a. A baby slept four hours and woke up to nurse. Then she slept another two hours and woke up to nurse. Then she slept three hours more and nursed again. Then she slept three hours until the morning.

How many hours did the baby sleep?

b. Mom needs 16 eggs to make cakes. The store sells eggs in cartons of 12. How many cartons does she need to buy?

How many eggs will she have left over?

6. Find the missing numbers. You can also work backwards, starting from 70!

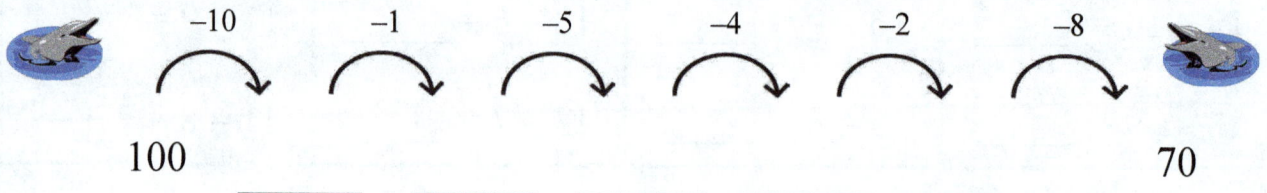

-10 -1 -5 -4 -2 -8

100 ___ ___ ___ ___ ___ 70

Review

1. Here are the 20 addition facts with single-digit numbers where the sum is between 10 and 20. Connect the problems to the right answer.

$5 + 6$		$4 + 8$			$6 + 9$
$6 + 8$	**11**	$6 + 7$	**15**		$8 + 8$
$6 + 6$	**12**	$9 + 4$	**16**		$7 + 8$
$4 + 7$	**13**	$7 + 7$	**17**		$9 + 8$
$3 + 9$		$2 + 9$			$7 + 9$
$3 + 8$	**14**	$5 + 7$	**18**		$9 + 9$
$8 + 5$		$5 + 9$			

2. Connect with a line the problems that are from the same fact family. You don't need to write the answers.

$13 - 7 = \square$	$12 - 5 = \square$	$15 - 7 = \square$
$7 + \square = 15$	$11 - 8 = \square$	$13 - 6 = \square$
$11 - 3 = \square$	$9 + \square = 17$	$5 + \square = 14$
$8 + \square = 17$	$15 - 8 = \square$	$17 - 8 = \square$
$14 - 5 = \square$	$6 + \square = 13$	$3 + \square = 11$
$7 + \square = 12$	$9 + \square = 14$	$\square + 5 = 12$

3. Find the differences.

a. The difference of 80 and 87 _____	**b.** The difference of 45 and 2 _____
c. The difference of 15 and 8 _____	**d.** The difference of 13 and 4 _____

4. Find the missing numbers.

a. $8 + \boxed{} = 15$	b. $7 + \boxed{} = 14$	c. $6 + \boxed{} = 13$
d. $13 - \boxed{} = 5$	e. $14 - \boxed{} = 8$	f. $15 - \boxed{} = 9$
g. $11 - 6 = \boxed{}$	h. $12 - 7 = \boxed{}$	i. $12 - 4 = \boxed{}$

5. Find the missing steps.

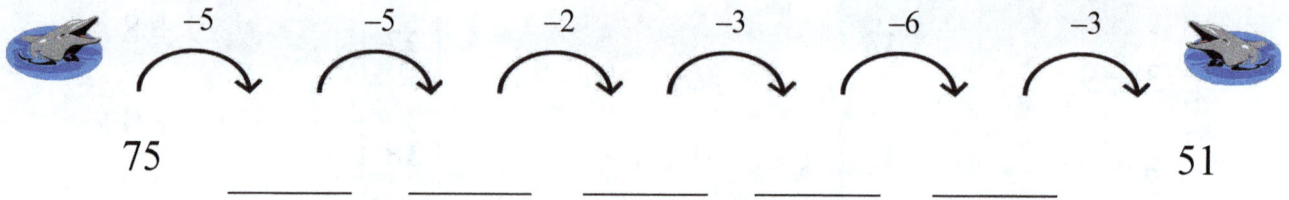

$$-5 \qquad -5 \qquad -2 \qquad -3 \qquad -6 \qquad -3$$

75 ___ ___ ___ ___ 51

6. a. You have an *odd* number of cookies and so does your friend. You put your cookies together and share them. Can you share them evenly or not?

Cookies you have	Cookies your friend has	Together we have	even/odd	Can you share evenly?
3	5			
5	9			
9	3			
9	7			

b. You have an *odd* number of cookies and your friend has an *even* number of cookies. You put your cookies together and share them. Can you share them evenly or not?

Cookies you have	Cookies your friend has	Together we have	even/odd	Can you share evenly?
5	6			
7	8			
9	4			
1	12			

7. Solve the puzzle. What happened to the teddy bear in the desert?

| 5 + 9 | 7 + 8 | | 13 − 8 | 2 + 9 | 10 + 5 | | 9 + 7 | 4 + 7 | 9 + 6 |

____ ____   ____ ____ ____   ____ ____ ____

| 7 + 7 | 13 − 6 | | 19 − 4 | 11 + 5 | 13 − 7 | | 3 + 13 | 11 − 5 | 13 − 4 | 6 + 9 |

____ ____   ____ ____ ____   ____ ____ ____ ____

Key:

A	E	I	O	G	H	T	W	N
9	6	14	11	5	16	15	8	7

8. Solve the word problems.

a. Jack has 13 tennis balls and Jane has 20.
How many more does Jane have than Jack?

b. Emma has three more flowers than Sofia. If Emma
has 14 flowers, how many does Sofia have?

c. In a chess game, Jacob has 2 more pawns than Anna.
If Anna has five pawns, how many does Jacob have?

d. You have $20, and you want to buy a Lego set that costs $28.
How many dollars do you still need to save?

Later, a neighbor pays you $2 for helping rake leaves.
How much more money do you need after that?

e. In a board game, you need to move 18 more squares to get to the end
of the game. You roll 6 and 5 on two dice and move that many squares.
Now how many more squares are there to the end?

What kind of numbers on the two dice would get you to the end?

Math Mammoth Add & Subtract 2-A
Answer Key

Add Using "Just One More", p. 9

1.

a. $8 + \underline{2} = 10$ $8 + \underline{3} = 11$	b. $4 + \underline{6} = 10$ $4 + \ 7 = 11$	c. $\underline{7} + 3 = 10$ $\underline{8} + 3 = 11$
d. $\underline{1} + 9 = 10$ $2 + 9 = 11$	e. $5 + \underline{5} = 10$ $5 + \ 6 = 11$	f. $\underline{4} + 4 = 8$ $5 + \ 4 = 9$

2. a. 3, 4   b. 2, 3   c. 4, 5   d. 6   e. 2   f. 8

3.

a.	b.	c.	d.
$7 + 2 = 9$	$5 + 6 = 11$	$4 + 6 = 10$	$2 + 9 = 11$
$3 + 8 = 11$	$3 + 4 = 7$	$2 + 8 = 10$	$5 + 4 = 9$
$5 + 5 = 10$	$6 + 4 = 10$	$7 + 4 = 11$	$3 + 7 = 10$

4. a. 13   b. 14   c. 17   d. 16   e. 11   f. 19   g. 15   h. 18   i. 11   j. 17   k. 13   l. 15

5. a. Joe gave away ten balloons. Joe still has two balloons.
   b. Marsha found thirteen shirts.
   c. She had to wash ten shirts.
   d. There were three more girls.
   e. There were just enough shirts for everyone.

A "Trick" with Nine and Eight, p. 11

1.

a. 9 + 6 $10 \quad + 5 \ = 15$	b. 9 + 4 $10 \quad + 3 \ = 13$
c. 9 + 3 $10 \quad + 2 \ = 12$	d. 9 + 5 $10 \quad + 4 \ = 14$

2.

a. $9 + 8$	b. $9 + 7$	c. $9 + 9$
$\diagup \ \diagdown$	$\diagup \ \diagdown$	$\diagup \ \diagdown$
$9 + \underline{1} + \underline{7}$	$9 + \underline{1} + 6$	$9 + \underline{1} + 8$
$10 + 7 = 17$	$10 + 6 = 16$	$10 + 8 = 18$

3.

a. 8 + 6 10 + _4_ = 14	b. 8 + 7 10 + _5_ = 15
c. 8 + 3 10 + _1_ = 11	d. 8 + 4 10 + _2_ = 12

4.

a. 8 + 8 / \ 8 + _2_ + 6 10 + 6 = 16	b. 8 + 5 / \ 8 + _2_ + 3 10 + 3 = 13	c. 8 + 7 / \ 8 + _2_ + 5 10 + 5 = 15

5. a. Not correct. 6 + 6 = 12   d. Not correct. 9 + 7 = 16

6. a. There are six apples left.   b. Jeremy has 15 apples.
   c. Jeremy picked two more flowers. Together, they have 16 flowers.
   d. The train of cars was 14 cm long.

7.

a. 7 + 3 + 5 = _15_	b. 9 + 1 + 2 = 12	c. 7 + 3 + 5 = 15
d. 6 + 4 + 6 = 16	e. 8 + 2 + 4 = 14	f. 5 + 5 + 8 = 18

8. a. 14 b. 15 c. 13 d. 12 e. 15 f. 18 g. 14 h. 16 i. 11

Puzzle corner: a. 8 b. 6 c. 4

Adding Within 20, p. 14

1.

a. 5 + 5 = 10	b. 6 + 6 = 12	c. 7 + 7 = 14
_5_ + _6_ = 11 and _6_ + _5_ = 11	6 + 7 = 13 and 7 + 6 = 13	7 + 8 = 15 and 8 + 7 = 15
d. 8 + 8 = 16	e. 9 + 9 = 18	f. 10 + 10 = 20
8 + 9 = 17 and 9 + 8 = 17	9 + 10 = 19 and 10 + 9 = 19	10 + 11 = 21 and 11 + 10 = 21

2. a. 17 b. 12 c. 14 d. 15

3.

a. 1 + 9 = 10	b. 3 + 7 = 10	c. 8 + 2 = 10
1 + 10 = 11 or 2 + 9 = 11	3 + 8 = 11 or 4 + 7 = 11	8 + 3 = 11 or 9 + 2 = 11
d. 6 + 4 = 10	e. 5 + 5 = 10	f. 7 + 3 = 10
6 + 5 = 11 or 7 + 4 = 11	5 + 6 = 11 or 6 + 5 = 11	7 + 4 = 11 or 8 + 3 = 11

4. a. $7 + 7 = 14$ Doubles chart
   b. $9 + 7 = 16$ Trick with nine
   c. $8 + 3 = 11$ Trick with eight
   d. $6 + 7 = 13$ Just one more than a double
   e. $5 + 6 = 11$ "Just one more" than a sum with 10
   f. $5 + 8 = 13$ Trick with eight
   g. $8 + 8 = 16$ Doubles chart
   h. $4 + 9 = 13$ Trick with nine

5. a. $9 + 5 - 2 = \$12$ Maria has twelve dollars.
   b. $9 + 3 = 12$; $8 + 3 = 11$ Ashley has one more shirt.
   c. $10 - 6 - 1 = 3$ Emily has three dollars.
   d. $8 - 2 + 4 = 10$ They have 10 tennis balls.

6.

a. $8 + 2 = 10$	b. $5 + 3 = 8$	c. $9 + 2 = 11$	d. $7 + 3 = 10$
$8 + 4 = 12$	$5 + 5 = 10$	$9 + 4 = 13$	$7 + 5 = 12$
$8 + 6 = 14$	$5 + 7 = 12$	$9 + 6 = 15$	$7 + 7 = 14$
$8 + 8 = 16$	$5 + 9 = 14$	$9 + 8 = 17$	$7 + 9 = 16$

7.

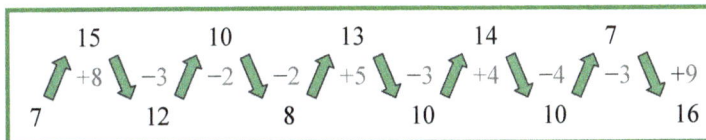

8. a. 18, 28, 38, 48, 58, 68, 78, 88, 98
   b. 27, 37, 47, 57, 67, 77, 87, 97, 107

9.

a. $6 + 6 = 12$	b. $8 + 8 = 16$	c. $6 + 5 = 11$
$6 + 7 = 13$	$9 + 7 = 16$	$7 + 4 = 11$

10. $9 + 6 = 15$ cm   Check the student's lines.

11.

a.	b.	c.	d.
$8 + 8 = 16$	$7 + 8 = 15$	$7 + 7 = 14$	$5 + 8 = 13$
$2 + 9 = 11$	$9 + 6 = 15$	$9 + 8 = 17$	$3 + 9 = 12$
$7 + 5 = 12$	$6 + 5 = 11$	$7 + 4 = 11$	$7 + 6 = 13$
e.	f.	g.	h.
$9 + 4 = 13$	$8 + 6 = 14$	$9 + 2 = 11$	$6 + 9 = 15$
$4 + 8 = 12$	$6 + 6 = 12$	$8 + 5 = 13$	$8 + 7 = 15$
$6 + 7 = 13$	$5 + 9 = 14$	$5 + 7 = 12$	$8 + 4 = 12$
i.	j.	k.	l.
$9 + 3 = 12$	$4 + 9 = 13$	$9 + 9 = 18$	$8 + 9 = 17$
$4 + 7 = 11$	$7 + 7 = 14$	$6 + 8 = 14$	$5 + 6 = 11$
$9 + 5 = 14$	$3 + 8 = 11$	$6 + 6 = 12$	$8 + 3 = 11$

Puzzle corner:

1. b. $14 - 4 = 10$ c. $16 - 6 = 10$ d. $15 - 5 = 10$

2. a. $13 - 3 = 10$ b. $17 - 7 = 10$ c. $19 - 9 = 10$

3.

a. $14 - 7$ $\diagup \ \diagdown$ $14 - 4 \ - 3$ $10 - 3 = \underline{7}$	b. $15 - 8$ $\diagup \ \diagdown$ $15 - \underline{5} - \underline{3}$ $10 - \underline{3} = \underline{7}$	c. $16 - 8$ $\diagup \ \diagdown$ $16 - \underline{6} \ - \underline{2}$ $10 - \underline{2} = \underline{8}$
d. $13 - 6$ $\diagup \ \diagdown$ $13 - \underline{3} - \underline{3}$ $10 - \underline{3} = \underline{7}$	e. $12 - 6$ $\diagup \ \diagdown$ $12 - \underline{2} - \underline{4}$ $10 - \underline{4} = \underline{6}$	f. $13 - 4$ $\diagup \ \diagdown$ $13 - \underline{3} - \underline{1}$ $10 - \underline{1} = \underline{9}$

4.

a. $12 - 6$ $\diagup \ \diagdown$ $12 - \underline{2} - 4 = \underline{6}$	b. $15 - 9$ $\diagup \ \diagdown$ $15 - \underline{5} - 4 = \underline{6}$	c. $13 - 8$ $\diagup \ \diagdown$ $13 - \underline{3} - \underline{5} = \underline{5}$
d. $13 - 7$ $\diagup \ \diagdown$ $13 - \underline{3} - \underline{4} = \underline{6}$	e. $14 - 7$ $\diagup \ \diagdown$ $14 - \underline{4} - \underline{3} = \underline{7}$	f. $12 - 4$ $\diagup \ \diagdown$ $12 - \underline{2} - \underline{2} = \underline{8}$

5. a. 7   b. 8   c. 7   d. 8   e. 7   f. 9   g. 8   h. 5

6. a. $13 - 8 = 5$ <u>Tom is five years older than Juan.</u>
   b. $13 - 9 = 4$ <u>Tom is four years older than Alice.</u>
   c. $15 - 10 = 5$ <u>Tom is still five years older than Juan.</u>

7.

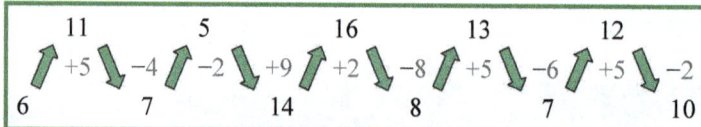

1.

a.	b.
$8 + 5 = 13$ $5 + 8 = 13$ $13 - 8 = 5$ $13 - 5 = 8$	$9 + 7 = 16$ $7 + 9 = 16$ $16 - 9 = 7$ $16 - 7 = 9$

2.

a. $8 + 4 = 12$ $12 - 8 = 4$ $12 - 4 = 8$	b. $9 + 7 = 16$ $16 - 9 = 7$ $16 - 7 = 9$	c. $7 + 6 = 13$ $13 - 7 = 6$ $13 - 6 = 7$

3.

a. $11 - 3 = 8$ $3 + 8 = 11$	b. $11 - 4 = 7$ $4 + 7 = 11$	c. $12 - 3 = 9$ $3 + 9 = 12$

4.

a. $14 - 8 = 6$ $8 + 6 = 14$	b. $15 - 7 = 8$ $7 + 8 = 15$	c. $17 - 8 = 9$ $8 + 9 = 17$
d. $12 - 8 = 4$ $8 + 4 = 12$	e. $16 - 7 = 9$ $7 + 9 = 16$	f. $13 - 7 = 6$ $7 + 6 = 13$
g. $13 - 8 = 5$ $8 + 5 = 13$	h. $11 - 7 = 4$ $7 + 4 = 11$	i. $14 - 9 = 5$ $9 + 5 = 14$

5.

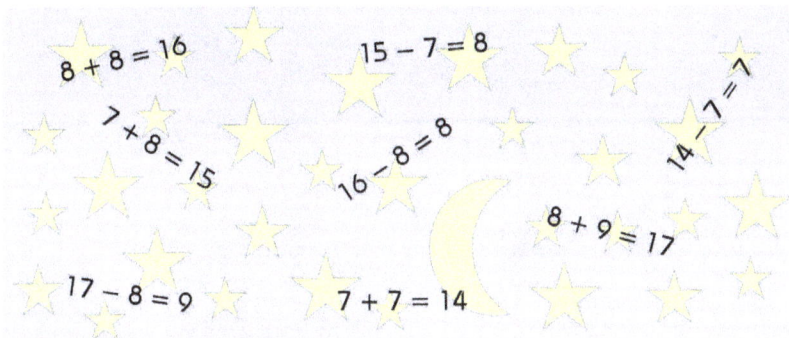

The matching additions and subtractions are:
$8 + 8 = 16$ and $16 - 8 = 8$
$7 + 8 = 15$ and $15 - 7 = 8$
$8 + 9 = 17$ and $17 - 8 = 9$
$7 + 7 = 14$ and $14 - 7 = 7$

6.

a. $12 - 8 = 4$	b. $11 - 7 = 4$	c. $13 - 9 = 4$
d. $15 - 6 = 9$	e. $18 - 9 = 9$	f. $16 - 7 = 9$

7. a. $15 - 6 = 9$ <u>Marsha has nine crayons.</u>
   $6 + 6 = 12$ <u>Susana has 12 crayons now.</u>
   <u>Susana has more crayons.</u>
   <u>Susana has three more crayons.</u>
   b. $7 + 8 = 15$. <u>She has 15 stars in her drawing.</u>
   c. $12 - 7 = 5$ <u>Matthew has $5.</u>
   d. $6 + 8 = 14$ <u>Neither are right. The total is 14.</u>

63

8.

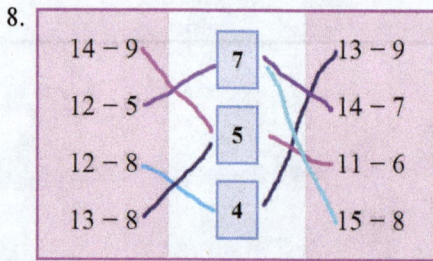

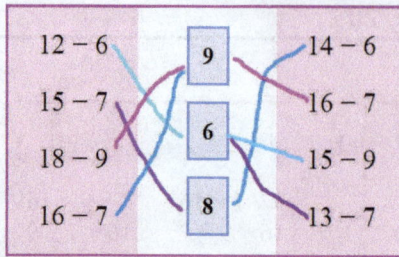

Review: Completing the Next Whole Ten, p. 23

1. a. 50, 56, **60**; d. 30, 37, **40**
   b. **70**, 72, 80; e. 20, 25, 30; middle of both
   c. **90**, 94, 100; f. **30**, 31, 40

2. a. 3, 3 b. 62 + 8 = 70, 2 + 8 = 10
   c. 94 + 6 = 100, 4 + 6 = 10

3. a. 42 + 8 = 50 b. 34 + 6 = 40 c. 66 + 4 = 70
   d. 61 + 9 = 70 e. 97 + 3 = 100 f. 83 + 7 = 90

4. The even numbers are 8, 12, 10, and 6.

5.

a. 8 = 4 + 4	b. 12 = 6 + 6
c. 10 = 5 + 5	d. 6 = 3 + 3

6. a. 7, 8   b. 5, 6   c. 6, 7

7.

13 + 6	54 − 6	73 + 8	45 + 7	99 + 4
15 + 9	14 + 8	15 + 5	13 + 6	32 + 7
45 + 7	73 − 7	64 + 5	82 + 9	16 + 7
30 + 12	39 − 1	74 + 6	73 + 9	52 + 7
46 + 7	32 + 7	31 + 9	86 − 4	65 + 4
92 + 4	21 + 8	24 + 7	22 − 8	32 + 6
83 + 6	11 + 7	98 + 2	57 − 3	17 + 9
44 + 9	12 − 8	95 + 6	38 + 5	53 + 9
71 + 9	34 + 4	36 + 7	19 + 4	28 + 11
53 + 7	29 + 2	26 + 6	78 + 6	32 + 5

8. a. 2, 4, 3   b. 3, 5, 1   c. 3, 3, 3

9. Answers will vary. For example:
   90 + 9 + 1 = 100; 90 + 8 + 2 = 100; 90 + 7 + 3 = 100;
   90 + 6 + 4 = 100; 90 + 5 + 5 = 100; 90 + 4 + 6 = 100;
   90 + 3 + 7 = 100; 90 + 2 + 8 = 100; 90 + 1 + 9 = 100

Review: Going Over Ten, p. 25

1.

a. 8 + 4 = 10 + 2 = 12	b. 9 + 5 = 10 + 4 = 14
c. 8 + 6 = 10 + 4 = 14	d. 9 + 3 = 10 + 2 = 12
e. 7 + 5 = 10 + 2 = 12	f. 9 + 8 = 10 + 7 = 17

2. a. (7 + _3_) + 2 = 12 b. (5 + _5_) + 3 = 13 c. (8 + _2_) + 4 = 14
   d. (6 + _4_) + 4 = 14 e. (9 + _1_) + 7 = 17 f. (7 + _3_) + 5 = 15

3.

a. 8 + 7
     / \\
8 + _2_ + _5_
 	
10 + _5_ = _15_
	b. 8 + 9
	     / \\
	8 + _2_ + _7_
	
	10 + _7_ = _17_
		c. 8 + 5
		     / \\
		8 + _2_ + _3_
		
		10 + _3_ = _13_

d. 9 + 4
     / \\
9 + _1_ + _3_
 	
10 + _3_ = _13_
	e. 9 + 6
	     / \\
	9 + _1_ + _5_
	
	10 + _5_ = _15_
		f. 9 + 9
		     / \\
		9 + _1_ + _8_
		
		10 + _8_ = _18_

4. a. 2, 3, 4   b. 3, 4, 5   c. 1, 2, 3   d. 4, 5, 6

5. The even numbers are 24, 58, 40, 100, and 2.

6. a. <u>You have $1 left.</u> $8 − $5 − $2 = $1   b. <u>She has $15.</u> $8 + $5 + $2 = $15   c. <u>Now, he has $7.</u> $8 − $3 + $2 = $7

Adding with 9, p. 27

1. a. 14, 14   b. 9 + 4 = 13; 10 + 3 = 13   c. 9 + 7 = 16; 10 + 6 = 16
   d. 9 + 6 = 15; 10 + 5 = 15   e. 9 + 8 = 17; 10 + 7 = 17   f. 9 + 9 = 18; 10 + 8 = 18

2. 4, 6, 8   10, 12, 14   16, 18, 20

3. a. 15, 15   b. 17, 10 + 7 = 17   c. 14, 10 + 4 = 14
   d. 16, 10 + 6 = 16   e. 18, 10 + 8 = 18   f. 12, 10 + 2 = 12

4.

9 + 0 = 9	9 + 5 = 14	9 + 9 = 18	9 + 4 = 13
9 + 3 = 12	9 + 6 = 15	9 + 1 = 10	9 + 10 = 19
9 + 7 = 16	9 + 8 = 17	9 + 2 = 11	

5. a. 13, 17, 14   b. 16, 13, 13   c. 12, 11, 18   d. 14, 17, 15

6. a. 4, 6   b. 7, 5   c. 8, 2

Puzzle corner. a. 56, 25   b. 64, 98   c. 72, 35

Adding with 8, p. 29

1. a. 13   b. 10 + 2 = 12   c. 8 + 7; 10 + 5 = 15   d. 8 + 6; 10 + 4 = 14   e. 8 + 8; 10 + 6 = 16 f. 8 + 3; 10 + 1 = 11

2. 4, 6, 8   10, 12, 14   16, 18, 20

8 + 0 = 8	8 + 5 = 13	8 + 8 = 16	8 + 9 = 17
8 + 3 = 11	8 + 7 = 15	8 + 1 = 9	8 + 4 = 12
8 + 10 = 18	8 + 1 = 9	8 + 6 = 14	8 + 2 = 10

3. a. 12, 14   b. 15, 13   c. 11, 17   d. 5, 7   e. 4, 8   f. 3, 6

4. a. <u>Jack ate 13.</u> 8 + 5 = 13   b. <u>Eight years older.</u> 13 − 5 = 8 or 5 + <u>8</u> = 13

5. a.	b.	c.
8 + 2 = 10	18 + 2 = 20	$\frac{1}{2}$ of 0 is 0.
8 + 4 = 12	18 + 4 = 22	$\frac{1}{2}$ of 2 is 1.
8 + 6 = 14	18 + 6 = 24	$\frac{1}{2}$ of 4 is 2.
8 + 8 = 16	18 + 8 = 26	$\frac{1}{2}$ of 6 is 3.
8 + 10 = 18	18 + 10 = 28	$\frac{1}{2}$ of 8 is 4.
8 + 12 = 20	18 + 12 = 30	$\frac{1}{2}$ of 10 is 5.
8 + 14 = 22	18 + 14 = 32	$\frac{1}{2}$ of 12 is 6.

Adding with 7, p. 31

We have already studied these addition facts:

7 + 8 = 15	8 + 7 = 15
7 + 9 = 16	9 + 7 = 16
7 + 10 = 17	10 + 7 = 17

These are the new facts with 7:

7 + 4 = 11	7 + 6 = 13
7 + 5 = 12	7 + 7 = 14

1. a. 12; 13   b. 14; 15   c. 16; 17   d. 18; 19   e. 10; 11   f. 8; 9

7 + 0 = 7	7 + 5 = 12	7 + 6 = 13	7 + 9 = 16
7 + 3 = 10	7 + 9 = 16	7 + 7 = 14	7 + 4 = 11
7 + 10 = 17	7 + 8 = 15	7 + 1 = 8	7 + 2 = 9

2. a. 11, 13, 12   b. 15, 17, 10   c. 7, 6, 8   d. 5, 9, 4   e. 14, 16, 15   f. 11, 16, 10   g. 5, 8, 9   h. 10, 3, 5

3.

+ 7

4	11
7	14
8	15
10	17
5	12
9	16

+ 8

3	11
6	14
5	13
7	15
2	10
4	12

+ 9

2	11
4	13
7	16
8	17
3	12
5	14

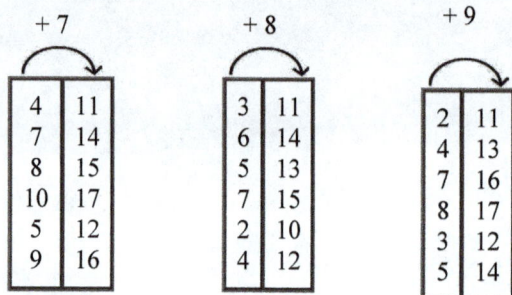

Adding with 6, p. 33

6 + 5 = 11	6 + 6 = 12

6 + 0 = 6	6 + 5 = 11	6 + 9 = 15	6 + 6 = 12
6 + 3 = 9	6 + 7 = 13	6 + 4 = 10	6 + 8 = 14
6 + 10 = 16	6 + 1 = 7	6 + 2 = 8	

1. a. 10, 12, 11   b. 14, 15, 13   c. 8, 10, 6   d. 6, 9, 5   e. 11, 13, 10   f. 15, 14, 12   g. 7, 6, 5   h. 7, 8, 9

2.

a.	b.	c.
1 + 6 + 9 = 16	3 + 6 + 7 + 2 = 18	6 + 5 + 1 + 4 = 16
6 + 8 + 2 = 16	1 + 5 + 5 + 7 = 18	8 + 3 + 2 + 6 = 19
5 + 7 + 5 = 17	2 + 7 + 8 + 2 = 19	9 + 6 + 1 + 4 = 20

3. a. 12 apples. 7 + 5 = 12   b. It cost $6. $6 + $6 = 12 or $12 − $6 = $6   c. Nine bananas. 4 + 1 + 2 + 2 = 9
   d. He solved 14 problems. 9 + 5 = 14   e. She solved 7 more problems than Ashley. 14 − 7 = 7 or 7 + 7 = 14

1.

2.

a.	b.	c.
9 + _10_ = 19	_1_ + 16 = 17	6 + _0_ = 6
8 + _10_ = 18	_3_ + 14 = 17	6 + _2_ = 8
7 + _10_ = 17	_5_ + 12 = 17	6 + _4_ = 10
_6_ + _10_ = _16_	_7_ + _10_ = _17_	_6_ + _6_ = _12_
_5_ + _10_ = _15_	_9_ + _8_ = _17_	_6_ + _8_ = _14_
_4_ + _10_ = _14_	_11_ + _6_ = _17_	_6_ + _10_ = _16_
_3_ + _10_ = _13_	_13_ + _4_ = _17_	_6_ + _12_ = _18_
_2_ + _10_ = _12_	_15_ + _2_ = _17_	_6_ + _14_ = _20_
_1_ + _10_ = _11_	_17_ + _0_ = _17_	_6_ + _16_ = _22_

3.

+	6	8	4	5	7	3	9
7	13	15	11	12	14	10	16
9	15	17	13	14	16	12	18
5	11	13	9	10	12	8	14

4. a. <u>18 elephants.</u> 4 + 14 = 18   b. <u>Annie has five dolls.</u> 10 − 5 = 5.
   c. <u>He emptied 9 baskets.</u> 5 + 4 = 9.   d. <u>She ate 17.</u> 10 + 7 = 17

5. a. 14, 14 b. 17, 12 c. 18, 18

Difference and How Many More, p. 37

1. a. 4   b. 4   c. 12   d. 11

2.

The difference between 10 and 4 a. $10 - 4 = 6$	The difference between 2 and 9 b. $9 - 2 = 7$	The difference between 8 and 3 c. $8 - 3 = 5$
The difference between 20 and 50 d. $50 - 20 = 30$	The difference between 10 and 90 e. $90 - 10 = 80$	The difference between 19 and 8 f. $19 - 8 = 11$

3. a. 4   b. 2   c. 4   d. 5   e. 1   f. 3   g. 5   h. 1

4.

a. The difference between 10 and 6 is _4_. $6 + \underline{4} = 10$	b. The difference between 6 and 12 is _6_. $6 + \underline{6} = 12$
c. The difference between 15 and 8 is _7_. $\underline{8} + \underline{7} = \underline{15}$	d. The difference between 4 and 11 is _7_. $\underline{4} + \underline{7} = \underline{11}$

5. a. 3   b. 2   c. 5

6. a. 4   b. 2   c. 34   d. 64   e. 20   f. 60

7. a. <u>Jen has read three more pages.</u> $20 - 17 = 3$ or $17 + 3 = 20$
   b. <u>Mom has 17 eggs.</u> $12 + 5 = 17$
   c. <u>She has eight pages left.</u> $42 + 8 = 50$ or $50 - 42 = 8$
   d. <u>Andy worked more hours. He worked three hours more.</u> Janet worked: $2 + 3 = 5$ hours. The difference: $8 - 5 = 3$.
   e. <u>She killed 23 more flies than her husband.</u> $28 - 5 = 23$
   f. <u>She killed 19 flies.</u> $5 + 12 + 2 = 19$
   g. <u>Matthew has more. He has $2 more.</u> Matthew has: $12 + $5 = 17. Bob has $6 + $9 = 15.

Number Rainbows - 11 and 12, p.40

The student is supposed to practice mentally and not write the answers down. The answers are of course easily seen from the rainbow. Encourage the child to also practice the subtraction problems while keeping the rainbow covered. There is a page for number rainbows with 13 and 14 also (a little later). Feel free to also draw or have the child draw (better!) ones for 15, 16, 17, and 18.

Fact Families with 11, p. 42

1.

Fact families with 11		
10, 1, and 11 	$10 + 1 = 11$ $1 + 10 = 11$	$11 - 10 = 1$ $11 - 1 = 10$
9, 2, and 11 	$9 + 2 = 11$ $2 + 9 = 11$	$11 - 2 = 9$ $11 - 9 = 2$
8, 3, and 11 	$8 + 3 = 11$ $3 + 8 = 11$	$11 - 8 = 3$ $11 - 3 = 8$
7, 4, and 11 	$7 + 4 = 11$ $4 + 7 = 11$	$11 - 4 = 7$ $11 - 7 = 4$
6, 5, and 11 	$6 + 5 = 11$ $5 + 6 = 11$	$11 - 6 = 5$ $11 - 5 = 6$

2.

a. $11 - 10 = 1$ $11 - 9 = 2$ $11 - 6 = 5$ $11 - 8 = 3$	b. $11 - 2 = 9$ $11 - 4 = 7$ $11 - 5 = 6$ $11 - 7 = 4$	c. $11 - 3 = 8$ $11 - 6 = 5$ $11 - 9 = 2$ $11 - 4 = 7$

Fact Families with 12, p. 43

1.

Fact families with 12		
10, 2, and 12	$10 + 2 = 12$ $2 + 10 = 12$	$12 - 10 = 2$ $12 - 2 = 10$
9, 3, and 12	$9 + 3 = 12$ $3 + 9 = 12$	$12 - 9 = 3$ $12 - 3 = 9$
8, 4, and 12	$8 + 4 = 12$ $4 + 8 = 12$	$12 - 8 = 4$ $12 - 4 = 8$
7, 5, and 12	$7 + 5 = 12$ $5 + 7 = 12$	$12 - 7 = 5$ $12 - 5 = 7$
6, 6, and 12	$6 + 6 = 12$	$12 - 6 = 6$

2.

a.	b.	c.	d.
$12 - 4 = 8$ $11 - 9 = 2$ $12 - 8 = 4$ $11 - 6 = 5$	$11 - 8 = 3$ $12 - 7 = 5$ $11 - 3 = 8$ $12 - 5 = 7$	$12 - 6 = 6$ $11 - 4 = 7$ $12 - 9 = 3$ $12 - 4 = 8$	$12 - 3 = 9$ $12 - 10 = 2$ $11 - 5 = 6$ $11 - 7 = 4$

3. a. 5, 3   b. 5, 4   c. 2, 4   d. 6, 3

4. Answers will vary. For example: Think of the addition $8 + \underline{\quad} = 11$.
   The number that is missing is 3 and that is the answer to $11 - 8$.

5.

a.	b.	c.
$16 - 1 = 15$	$0 + 17 = 17$	$15 - 1 = 14$
$16 - 3 = 13$	$2 + 15 = 17$	$15 - 3 = 12$
$16 - 5 = 11$	$4 + 13 = 17$	$15 - 5 = 10$
$16 - 7 = 9$	$6 + 11 = 17$	$15 - 7 = 8$
$16 - 9 = 7$	$8 + 9 = 17$	$15 - 9 = 6$
$16 - 11 = 5$	$10 + 7 = 17$	$15 - 11 = 4$
$16 - 13 = 3$	$12 + 5 = 17$	$15 - 13 = 2$
$16 - 15 = 1$	$14 + 3 = 17$	$15 - 15 = 0$

Puzzle Corner:

a. $6 + 4 + 4 = 14$ $8 + 2 + 3 = 13$	b. $2 + 4 + 2 = 8$ $3 + 3 + 3 = 9$	c. $10 + 3 + 4 = 17$ $10 + 3 + 2 = 15$

Number Rainbows—13 and 14, p. 45

1.

0 1 2 3 4 5 6 7 8 9 10 11 12 13

0 1 2 3 4 5 6 7 8 9 10 11 12 13 14

7

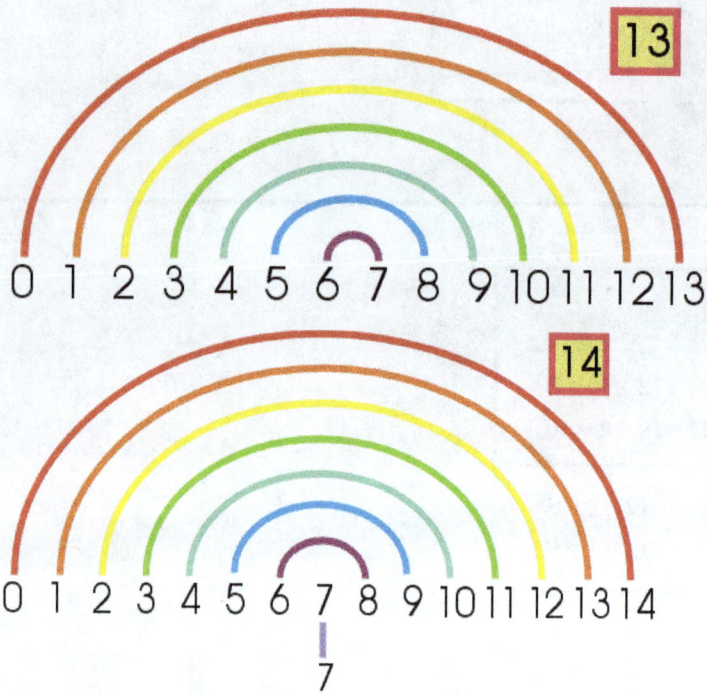

Fact Families - 13 and 14, p. 46

1.

Fact families with 13		
10, 3, and 13	$10 + 3 = 13$ $3 + 10 = 13$	$13 - 10 = 3$ $13 - 3 = 10$
9, 4, and 13	$9 + 4 = 13$ $4 + 9 = 13$	$13 - 9 = 4$ $13 - 4 = 9$
8, 5, and 13	$8 + 5 = 13$ $5 + 8 = 13$	$13 - 8 = 5$ $13 - 5 = 8$
7, 6, and 13	$7 + 6 = 13$ $6 + 7 = 13$	$13 - 7 = 6$ $13 - 6 = 7$

2.

$13 - 7 = \blacksquare$	$11 - 4 = \blacksquare$	$12 - 7 = \blacksquare$
$5 + \blacksquare = 12$	$11 - 8 = \blacksquare$	$13 - 6 = \blacksquare$
$11 - 3 = \blacksquare$	$5 + \blacksquare = 13$	$3 + \blacksquare = 12$
$8 + \blacksquare = 13$	$12 - 5 = \blacksquare$	$13 - 5 = \blacksquare$
$12 - 3 = \blacksquare$	$6 + \blacksquare = 13$	$3 + \blacksquare = 11$
$7 + \blacksquare = 11$	$9 + \blacksquare = 12$	$4 + \blacksquare = 11$

3.

Fact families with 14		
10, 4, and 14	$10 + 4 = 14$ $4 + 10 = 14$	$14 - 10 = 4$ $14 - 4 = 10$
9, 5, and 14	$9 + 5 = 14$ $5 + 9 = 14$	$14 - 5 = 9$ $14 - 9 = 5$
8, 6, and 14	$8 + 6 = 14$ $6 + 8 = 14$	$14 - 8 = 6$ $14 - 6 = 8$
7, 7, and 14	$7 + 7 = 14$	$14 - 7 = 7$

4. a. 5, 8   b. 8, 9   c. 5, 6   d. 3, 5

5. a. 5   b. 8   c. 6   d. 13   e. 14   f. 12   g. 6   h. 5   i. 5

6. a. 18 cars.   b. 6 more.   c. 11 cherries.   d. 16 apples.

7. a. $40 + 8 = 48 + 8 = 56 + 8 = 64 + 8 = 72 + 8 = 80 + 8 = 88 + 8 = 96 + 8 = 104$
   b. $17 + 4 = 21 + 4 = 25 + 4 = 29 + 4 = 33 + 4 = 37 + 4 = 41 + 4 = 45 + 4 = 49$

Fact Families with 15, p. 49

1.

Fact families with 15		
10, 5, and 15	$10 + 5 = 15$ $5 + 10 = 15$	$15 - 10 = 5$ $15 - 5 = 10$
9, 6, and 15	$9 + 6 = 15$ $6 + 9 = 15$	$15 - 6 = 9$ $15 - 9 = 6$
8, 7, and 15	$8 + 7 = 15$ $7 + 8 = 15$	$15 - 8 = 7$ $15 - 7 = 8$

2. a. 10   b. 7   c. 11   d. 6   e. 9   g. 8

3. Answers will vary: For example, think of the addition $9 + \_\_\_ = 15$. Because $9 + 6 = 15$, then $15 - 9 = 6$.

4. 9, 12, 15, 18, 21, 24, 27, 30, 33

5. a. Jacob has 10 peaches. $7 + 3 = 10$.   b. Ethan has 8 books. $11 - 3 = 8$, or $8 + \_\_\_ = 11$.
   c. Noah picked 6 more. $15 - 9 = 6$ or $9 + \_\_\_ = 15$.   d. Sophia picked 10. $15 - 5 = 10$.

6. a. $6 = 3 + 3$	b. $12 = 6 + 6$	c. $10 = 5 + 5$
d. $18 = 9 + 9$	e. $20 = 10 + 10$	f. $8 = 4 + 4$

7. They have 8 left. $7 + 9 = 16$. Half of 16 is 8.

8. 0, 2, 4, 6, 8, 10, 12, 14, 16, 18, 20

9. a. bike, $28, and kite, $30 together $ 58	b. jeans, $47, shoes, $30, and toy $10 together $ 87

Fact Families with 16, p. 51

1.

Fact families with 16		
10, 6, and 16	$10 + 6 = 16$ $16 - 10 = 6$	
	$6 + 10 = 16$ $16 - 6 = 10$	
9, 7, and 16	$9 + 7 = 16$ $16 - 9 = 7$	
	$7 + 9 = 16$ $16 - 7 = 9$	
8, 8, and 16	$8 + 8 = 16$ $16 - 8 = 8$	
	$8 + 8 = 16$ $16 - 8 = 8$	

2. a. 5, 3, 6   b. 4, 7, 5   c. 6, 5, 8   d. 8, 9, 6

3.

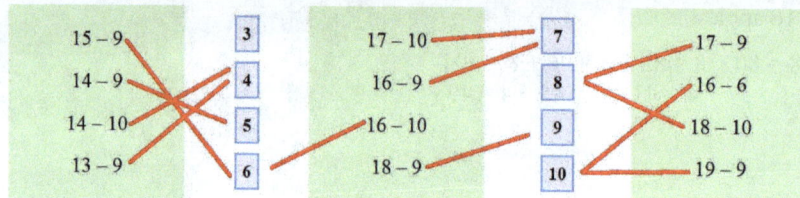

4. a. $6 + 3 = 9 + 3 = 12 + 3 = 15 + 3 = 18 + 3 = 21 + 3 = 24 + 3 = 27 + 3 = 30$
   b. $12 + 4 = 16 + 4 = 20 + 4 = 24 + 4 = 28 + 4 = 32 + 4 = 36 + 4 = 40 + 4 = 44$

5. a. <u>20 children.</u> $24 - 2 - 2 = 20$ b. **Yes.** $\$10 + \$4 = \$14$ c. $\$3. \$20 - \$17 = \$3$ d. <u>$4 more.</u> $\$12 + \_\_\_ = \$16$
   e. <u>More girls. There is 1 more girl than boys.</u> There are 5 boys and $7 - 1 = 6$ girls.

6. a. >   b. <   c. >   d. >   e. >   f. <

Fact Families - 17 and 18, p. 53

1.

Fact families with 17		
10, 7, and 17	$10 + 7 = 17$ $17 - 10 = 7$	
	$7 + 10 = 17$ $17 - 7 = 10$	
9, 8, and 17	$9 + 8 = 17$ $17 - 8 = 9$	
	$8 + 9 = 17$ $17 - 9 = 8$	

Fact families with 18		
10, 8, and 18	$10 + 8 = 18$ $18 - 8 = 10$	
	$8 + 10 = 18$ $18 - 10 = 8$	
9, 9, and 18	$9 + 9 = 18$ $18 - 9 = 9$	
	$9 + 9 = 18$ $18 - 9 = 9$	

2. a. 7, 8, 8, 9 b. 6, 7, 7, 8 c. 8, 7, 7, 6 d. 3, 4, 2, 3

3. a. > b. > c. < d. > e. < f. =

4. a. 6 b. 8 c. 9 d. 15 e. 15 f. 12 g. 8 h. 9 i. 9

5. a. <u>The baby slept 12 hours.</u> 4 + 2 + 3 + 3 = 12
   b. <u>Mom needs to buy two cartons of eggs.</u> 12 + 12 = 24, which is more than 16.
      <u>She will have eight eggs left.</u> 24 − 16 = 8

6. 100 - 90 - 89 - 84 - 80 - 78 - 70

Review, p. 55

1.
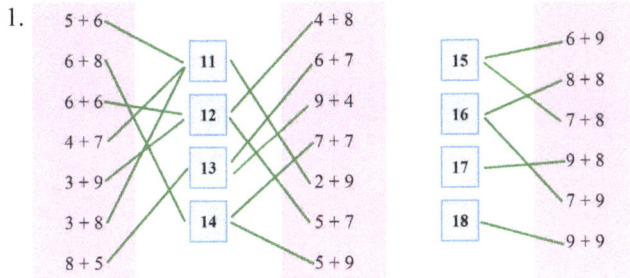

2.
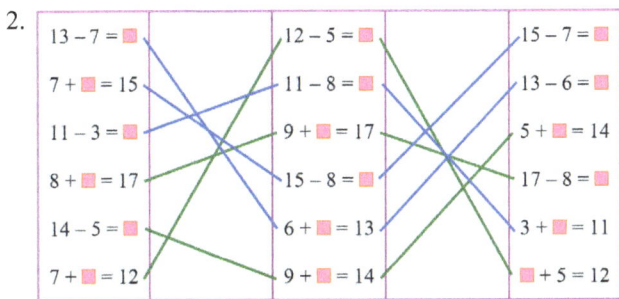

3. a. 7 b. 43 c. 7 d. 9

4. a. 7 b. 7 c. 7 d. 8 e. 6 f. 6 g. 5 h. 5 i. 8

5. 75, 70, 65, 63, 60, 54, 51

6. a.

Cookies you have	Cookies your friend has	Together we have	even/odd	Can you share evenly?
3	5	8	even	yes
5	9	14	even	yes
9	3	12	even	yes
9	7	16	even	yes

6. b.

Cookies you have	Cookies your friend has	Together we have	even/odd	Can you share evenly?
5	6	11	odd	no
7	8	15	odd	no
9	4	13	odd	no
1	12	13	odd	no

7. IT GOT HOT IN THE HEAT.

8. a. <u>Jane has seven more than Jack.</u> 20 − 13 = 7 or 13 + ___ = 20 b. <u>Sofia has 11.</u> 14 − 3 = 11
   c. <u>Jacob has 7 pawns.</u> 5 + 2 = 7 d. <u>You will need to save $8 more.</u> Think: $20 + ___ = $28.
      <u>After the neighbor pays you, you still need $6.</u> You have $20 + $2 = $22. Think: $22 + _6_ = $28.
   e. <u>I need seven more squares to get to the end of the game.</u> You roll 5 + 6 = 11, and 11 + _7_ = 18.
      <u>To get to the end, you need to roll seven on two dice.</u> You could roll 3 and 4, or 1 and 6, or 2 and 5.

More from math MAMMOTH

Math Mammoth has a variety of resources to fit your needs. All are available as economical downloads, and most also as printed copies.

- **Math Mammoth Light Blue Series**
  A complete curriculum for grades 1-7. Each grade level includes two student worktexts (A and B), which contain all the instruction and exercises all in the same book, answer keys, tests, cumulative reviews, and a worksheet maker. International (all metric), Canadian, and South African versions are also available.
  https://www.MathMammoth.com/complete-curriculum

  https://www.MathMammoth.com/international/international

  https://www.MathMammoth.com/canada/

  https://www.MathMammoth.com/south_africa/

- **Math Mammoth Skills Review Workbooks**
  These workbooks are intended to be used alongside the Light Blue series full curriculum, and they provide additional review to the topics studied in the main curriculum, in a spiral manner.
  https://www.MathMammoth.com/skills_review_workbooks/

- **Math Mammoth Blue Series**
  Blue Series books are topical worktexts for grades 1-7, containing both instruction and exercises. The topics cover all elementary mathematics from 1st through 7th grade. These books are not tied to grade levels, and are thus great for filling in gaps.
  https://www.MathMammoth.com/blue-series

- **Make It Real Learning**
  These activity workbooks concentrate on answering the question, "Where is math used in real life?" The series includes various workbooks for grades 3-12.
  https://www.MathMammoth.com/worksheets/mirl/

- **Review Workbooks**
  Workbooks for grades 1-7 that provide a comprehensive review of one grade level of math—for example, for review during school break or summer vacation.
  https://www.MathMammoth.com/review_workbooks/

Free gift!

- Receive over 350 free sample pages and worksheets from my books, plus other freebies:
  https://www.MathMammoth.com/worksheets/free

Lastly...

- Inspire4 is an inspirational website for the whole family I've been privileged to help with:
  https://www.inspire4.com